SpringerBriefs in GIS

SpringerBriefs in GIS present compact, concise summaries of cutting-edge research, practical applications and visualizations in the use of geographical information systems. At 50 to 125 pages (approx. 20,000–50,000) words, SpringerBriefs in GIS provides researchers and practitioners with an innovative venue to present work that might be longer and more complex in scope than a journal article.

This series aims to cover a wide range of topics related to geographical information science and geographical information systems. Potential topics could include: an in-depth case study on the use of GIS to accomplish a specific goal; a guide to an emerging GIS tool, technique or map; a "hot-take" on or snapshot of a current issue that needs to be published as quickly as possible (just 8–12 weeks after a manuscript's delivery and acceptance). Multidisciplinary studies are particularly welcome.

SpringerBriefs are distributed through the same channels as Springer's book content, and are available as physical books and full and chapter-wise eBooks. Both solicited and unsolicited manuscripts are considered for publication in this series. Please send questions and proposals to Zachary Romano, Associate Editor, Earth Science, Environment, and Geography, at Zachary.Romano@Springer.com.

More information about this series at http://www.springer.com/series/15244

Rupesh Jayaram Patil

Spatial Techniques for Soil Erosion Estimation

Remote Sensing and GIS Approach

 Springer

Rupesh Jayaram Patil
Department of Soil and Water Engineering,
  College of Agricultural Engineering
Jawaharlal Nehru Agricultural University
Jabalpur, Madhya Pradesh
India

ISSN 2367-010X   ISSN 2367-0118   (electronic)
SpringerBriefs in GIS
ISBN 978-3-319-74285-4   ISBN 978-3-319-74286-1   (eBook)
https://doi.org/10.1007/978-3-319-74286-1

Library of Congress Control Number: 2017964240

This Springer imprint is published by the registered company Springer International Publishing AG part of Springer Nature
The registered company address is: Gewerbestrasse 11, 6330 Cham, Switzerland

*To my parents Shri. Jayaram Patil
and Smt. Jayashree*

Foreword

It is my pleasure to write the foreword for this book entitled, "Spatial Techniques for Soil Erosion Estimation—Remote Sensing and GIS Approach." Soil erosion has been one of the main concerns of the developing world because of the increasing demographic pressure on natural resources and the threat of the erosional processes to these resources. This work is a significant contribution to the ongoing research on quantifying the rates of soil erosion. The use of Remote Sensing and Geographical Information Systems in distributed soil erosion modeling is a real advancement in this field. Although this thesis discusses all the basic approaches and some ongoing research to identify the extent of soil erosion, the major contribution of this work is the use of C++ programming in processing and analyzing the Digital Elevation Model (DEM). It provides valuable information on the erosional and depositional features of the area based on the topographic input. It makes this work different and in some sense a unique piece of research over the traditional analysis of soil erosion. The study was conducted in the Shakkar River watershed which has a long historical background and where the effect of fluvial processes on the formation and degradation of the landscape can be carefully observed. I hope the readers/researchers will find this work interesting and unique and may get some new ideas to put forth. I wish the author all the success in getting his thesis work published and wish him best of luck in his future endeavor.

Jabalpur, India

Dr. Shailesh Kumar Sharma
Professor
Department of Soil and Water Engineering
College of Agricultural Engineering
Jawaharlal Nehru Agricultural University

The original version of the book was revised: Copyright text has been added and acknowledgement has been updated. The erratum to the book is available at https://doi.org/10.1007/978-3-319-74286-1_7

Acknowledgements

I would like to express profound and heartfelt gratitude to my research supervisor Dr. S. K. Sharma for his invaluable advice, continued guidance, constructive encouragement, and sound support throughout the period of this research project. I am grateful to Dr. V. S. Tomar, Hon'ble Vice-Chancellor, and Dr. P. K. Mishra, Director of Instruction, Jawaharlal Nehru Krishi Vishwa Vidyalaya (Agricultural University), Jabalpur (MP) for allowing me to undertake such a challenging and intellectually demanding work. I would like to express my deep sense of gratitude to Dr. T. K. Bhattacharya, Dean, College of Agricultural Engineering, JNKVV, Jabalpur (MP) for his kind cooperation and words of inspiration when needed most during this research work. I am thankful to Dr. Deva Kant, Dr. M. K. Awasthi, and Dr. R. K. Nema from Department of Soil and Water Engineering, College of Agricultural Engineering, Jabalpur (MP) for their expert opinions and valuable suggestions. Thanks also goes to Dr. K. S. Kushwaha from Department of Mathematics and Statistics, College of Agricultural Engineering, Jabalpur (MP) for his valuable suggestions and guidance. I am very thankful to Subhash Thakur, Renu Upadhyay and Seema Suraiya from the College of Agricultural Engineering, Jabalpur (MP) for their kind cooperation. I am also thankful to Shiv Kumar Yadav for his help during the research period. I would like to extend thanks to my colleagues, Ashish Pitre, Kumar Soni, Rajiv Ranjan, Swapnil Ganvir, Deepak Chouhan and Ajeet Kumar for their encouragement and cooperation. Special thanks goes to Christopher Bryant. Special thanks also due to Dr. Michael Leuchner and Naomi Portnoy for their kind support and cooperation. I hereby acknowledge that several figures and tables in this book were previously published in the article: 'Use of remote sensing, GIS and C++ for soil erosion assessment in the Shakkar River basin, India' by Rupesh Jayaram Patil, Shailesh Kumar Sharma, Sanjay Tignath & Aribam Priya Mahanta Sharma, Hydrological Sciences Journal Vol 62:2 pp. 217–231 (2017), copyright © IAHS Press. I am grateful to Taylor & Francis Ltd, on behalf of IAHS Press, and all my co-authors for giving me permission to reprint them here.

I would like to thank my parents, Jayaram Patil and Jayashree Patil, my brother Rohit and other family members without whose encouragement and moral support I could not have achieved this stage. I owe much to all who have directly and indirectly contributed to the success of this research project work.

Jabalpur, India
June 2013
Rupesh Jayaram Patil

Reference

Patil RJ, Sharma SK, Tignath S, Sharma APM (2017) Use of remote sensing, GIS and C++ for soil erosion assessment in the Shakkar River basin, India. Hydrol Sci J 62(2):217–231

Contents

Abbreviations

A	Average annual rate of soil loss
C	Crop/cover management factor
CAE	College of Agricultural Engineering
cm	centimeter
CWC	Central Water Commission
DEM	Digital Elevation Model
E	East
EI_{30}	Rainfall erosion index unit
eq	equation
ERDAS	Earth Resources Data Analysis Systems
et al.	and others
etc.	Etcetera (and so on)
FAO	Food and Agricultural Organization
FCC	False Color Composite
Fig.	Figure
ft	Foot
GIS	Geographic Information Systems
GLCF	Global Land Cover Facility
h	Hour
ha	hectare
I	Rainfall intensity
i.e.	id est (that is)
IRS	Indian Remote Sensing Satellite
JNKVV	Jawaharlal Nehru Krishi Vishwa Vidyalaya
K	Soil erodibility factor
KE	Kinetic energy
km	Kilometer
L	Slope length factor
LISS	Linear Imaging Self-Scanner
m	meter

Mha	Million hectares
MJ	Mega Jules
mm	millimeter
MP	Madhya Pradesh
MPWSRP	Madhya Pradesh Water Sector Restructuring Project
MSL	Mean Sea Level
N	North
NBSSLUP	National Bureau of Soil Survey & Land Use Planning
NDVI	Normalized Difference Vegetation Index
NRSA	National Remote Sensing Agency
P	Conservation/support practice factor
pixel	Picture element
R	Rainfall erosivity factor
R^2	Coefficient of determination
RS	Remote Sensing
S	Slope gradient factor
SRTM	Shuttle Radar Topographic Mission
t	tones
TIFF	Tagged Information File Format
TIN	Triangular Irregular Network
USLE	Universal Soil Loss Equation
VI	Vegetation Index
yr	Year
°	degree
′	minute
%	percentage

List of Figures

List of Tables

Chapter 1
Introduction

Abstract Soil erosion is a global threat to the natural resources and is particularly responsible for reduction in crop yield due to reduction in land productivity and storage capacity of multipurpose reservoirs due to continuous siltation. Accelerated soil erosion has adverse economic and ecological impacts. The assessment of the risk of soil erosion can be helpful for land evaluation in the region where soil erosion is the major threat to sustained agriculture, as the soil is the basis of agricultural production. Erosion models are used to predict soil erosion. Soil erosion modeling can consider many of the complex interactions that influence rates of erosion by simulating erosion processes in the watershed. Most of these models need information related to soil type, land use, landform, climate and topography to estimate soil loss. One of the biggest problems in testing these models is the generation of input data, which is too spatial, and conventional methods proved to be too costly and time consuming for generating this input data. Due to advances in Remote Sensing technology, deriving the spatial information on input parameters has become more convenient and cost-effective. With the powerful spatial data processing capabilities of Geographic Information System (GIS) and its compatibility with RS data, soil erosion modeling approaches have become more comprehensive and robust.

Keywords Soil erosion · Remote Sensing · GIS · USLE

Soil erosion by water occurs throughout the world. It is the process of detachment or entrainment, transportation of surface soil particles from original location and deposition of it to a new depositional area. Various human activities disturb the land surface and thereby induce significant alteration to the natural rates of erosion. Water erosion has been recognized as the most severe hazard threatening the protection of soil as it reduces soil productivity by removing the most fertile topsoil (FAO/UNEP 1994). On the world map of the status of human-induced soil degradation, it is believed that the loss of topsoil and terrain deformation due to soil erosion are the consequences of deforestation, removal of natural vegetation, and overgrazing in the mountainous regions (Shrestha 1997).

R. J. Patil, *Spatial Techniques for Soil Erosion Estimation*, SpringerBriefs in GIS,
https://doi.org/10.1007/978-3-319-74286-1_1

In India, 69 Mha of land is critically degraded and 106 Mha is severely eroded at an average rate of 16.4 t/ha/year of soil detachment (Singh 2000). The country shares only 2% of world's geographical area but supports over 17.5% of the world's population. Thus, India's land resources are under immense pressure to fulfill the demands of increasing population. Therefore, soil erosion in India is one of the prime concerns of the nation as agriculture is being adversely affected. Declining land availability for agriculture, which is expected to be only 0.15 ha per capita by 2035 AD, shows the severity of the problem (Singh 2000).

The soil is a nonrenewable resource and sediment is largely responsible for compromising the amount and quality of the water and silting the water bodies (Desilva et al. 2010). Apart from the serious loss of production due to soil erosion and land degradation because of human activities and interferences, the other immediate reaction of this problem is seen through excessive and premature siltation of multipurpose reservoirs. Nearly 10% of the soil detached annually is being deposited in surface reservoirs resulting in a loss of about 2% of storage capacity annually. These reservoirs are constructed in India for various purposes, such as hydroelectric power generation, irrigation, domestic water supply, soil erosion control and flood mitigation involving high expenditure. Most of them have been designed to last at least 100 years, but the excessive siltation due to accelerated erosion has considerably decreased the useful life of these structures. About 6000 MT of soil is eroded every year in India from 80 Mha of cultivable land losing 8.4 metric tons of nutrients. The nutrient losses in this way are much greater than the quantity that is presently being used in the country (Singh 2000). Out of 16 rivers of the world, which experience severe erosion and carry heavy sediment load, three rivers namely Ganga, Brahmaputra, and Kosi occupy second, third, and fourth positions, respectively. Also, a rise in the amount spent on drought and flood relief programs annually is an indication of the land degradation in the country (Mishra et al. 2003).

1.1 Soil Erosion Threat to Environment and Agriculture

Soil erosion is a global environmental crisis that today threatens our natural environment and agriculture. Accelerated soil erosion has adverse economic and ecological impacts (Lal 1998). It creates on-site and off-site effects on productivity due to a decline in land/soil quality (Lal 2001). The current rate of agricultural land degradation worldwide by soil erosion and other factors is leading to an irreparable loss in productivity on about 6 million hectares of fertile land a year (Dudal 1981). Asia has the highest soil erosion rate of 74 t/acre/year, and Asian rivers contribute about 80% of the total sediments delivered to the world's oceans (El-Swaify 1997).

Soil resource is valuable for the livelihood of human beings. Sustainable use of land depends on conservation and potential use of soil and water resources. Soil erosion has long-term impacts as it causes the loss of fertile topsoil and reduces the productive capacity of the land and thereby creates a risk to global food security (Mosbahi et al. 2013). Landslide, mudslides, the collapse of man-made terraces, soil loss from steep slopes, and decline of forest/pasture areas are the main reasons for land resource degradation. It has been observed that the loss of fertile topsoil due to surface and gully erosion is a common phenomenon, and agricultural land has expanded to areas having marginal soil cover. Thus, natural resources in mountainous terrain are profoundly affected by land degradation because of intensive deforestation, overgrazing and subsistence agriculture due to population pressure, large-scale road construction, and mining along with other anthropogenic activities. Because of deforestation coupled with the influence of high rainfall, fragile terrains with steep slopes have become prone to severe soil erosion (El-Swaify 1997).

1.2 Need to Assess Soil Erosion Risk

The assessment of the risk of soil erosion can be helpful for land evaluation in the region where soil erosion is the major threat to sustained agriculture. Soil erosion assessment and mapping of erosion-prone areas serve the knowledge for soil conservation and watershed management. Therefore, the need has arisen to quantify the rates of erosion as these are of paramount importance to policy makers. Consequently, monitoring is essential to check the deterioration of the natural resources, particularly those affecting agricultural practices. Hence, it is crucial to assess soil erosion risk for soil conservation program.

Various methodologies to assess rates of soil erosion are available. Conventional techniques of field study for measurement of soil erosion and runoff are expensive and time consuming. Moreover, these techniques have limitations because of the complexity of the interactions and difficulty of generalizing from the results. The accuracy is likely to be poor, and it is not known where the soil came from and when (Hudson 1995). In the suspended sediment measurement, the discharge and sediment yield passing a point at the outlet of a catchment is monitored for quantitative indication of the amount of soil lost from the catchment. However, the origin and extent of the sediment load from the catchment itself remain unknown. Similarly, the eroded soil deposited at other locations without reaching the gauging station cannot be computed. Use of these experimental outputs for quantification of soil loss from a wider area with a diversity of situational variation is still debatable. Land surveying using reconnaissance methods for soil loss estimation considers the approximation of the amount of erosion by measuring the change of surface level. Nevertheless, it has limitation for arable land because the surface level is affected by cultivation and settlement.

The hydrological studies including frequency analysis are vital in practice. The geomorphological parameters directly or indirectly reflect almost the entire watershed-based causative factors affecting the rate of runoff and sediment loss. The surface features are the fundamental units of analysis before adopting any sophisticated tools to monitor watershed responses in connection to any of the hydrologic processes acting on it. The need for accurate information on runoff and sediment yield has grown rapidly over the past few decades along with the acceleration in watershed management programs for conservation, development, and beneficial use of these resources (Hakim and Chandrakaran 2005).

1.3 Soil Erosion Modeling

Erosion models are used to predict soil erosion. Soil erosion modeling can consider many of the complex interactions that influence rates of erosion by simulating erosion processes in the watershed. Various parametric models such as empirical (statistical/metric), conceptual (semiempirical), and physical process based (deterministic) models are available to compute soil loss. In general, these models are categorized depending on the physical processes simulated by the model, the model algorithms describing these processes, and the data dependence of the model. Empirical models are the simplest of all three model types. They are statistical in nature, based primarily on the analysis of observations, and seek to characterize response from these data. The data requirements for such models are usually less as compared to conceptual and physical based models. Conceptual models play an intermediary role between empirical and physical process based models. Physical process based models take into account the combination of the individual components that affect erosion, including the complex interactions between various factors and their spatial and temporal variability. These models are comparatively over-parameterized.

Most of these models need information related to soil type, land use, landform, climate, and topography to estimate soil loss. They are designed for a specific set of conditions concerning the particular area. The Universal Soil Loss Equation (USLE) (Wischmeier and Smith 1965) was designed to predict soil loss from sheet and rill erosion in specific conditions from agricultural fields. Modified Universal Soil Loss Equation (MUSLE) (Williams and Berndt 1972), a modified version of USLE, applies to other conditions by introducing a hydrological runoff factor for sediment yield estimation. The Water Erosion Prediction Project (WEPP) (Nearing and Lane 1989) is a process-based continuous simulation model developed to replace USLE (Okoth 2003). Areal Nonpoint Source Watershed Environment Response Simulation (ANSWERS) (Beasley et al. 1980) was designed to compute soil erosion within a watershed. The European Soil Erosion Model (EUROSEM) is a single process based model for assessing and predicting the risk of soil erosion in fields and small catchments. The Morgan, Morgan, and Finney (MMF) model is an empirical model developed for mean annual soil loss estimation from field-sized areas on hill slopes (Morgan et al. 1984, 1999) having a strong physical base.

The USLE (Wischmeier and Smith 1965) is the simplest mathematical model, which has been used worldwide since the 1960s. The USLE is an empirical model, which serves to estimate annual soil loss. To obtain quick and efficient information regarding soil loss, a GIS–USLE integration is often used (Fistikoglu and Harmancioglu 2002). The USLE estimates long-term average soil losses from agricultural watersheds. The USLE is also likely to remain the simplest method available to undertake soil erosion studies on large basins (Julien and Tanago 1991). Extensive use of the USLE at a larger scale has been reported by many researchers (Jain and Kothyari 2000; Jain and Goel 2002; Pradhan et al. 2012; Khosrokhani and Pradhan 2014).

1.4 Remote Sensing and GIS in Soil Erosion Assessment

One of the biggest problems in testing these models is the generation of input data, which is very spatial. Conventional methods proved to be very costly and time consuming for generating this input data. With the advent of Remote Sensing (RS) technology, deriving the spatial information on input parameters has become more feasible and cost-effective. Besides, with the powerful spatial data processing capabilities of Geographic Information System (GIS) and its compatibility with RS data, soil erosion modeling approaches have become more comprehensive and robust (Bhaware 2006). RS can facilitate studying the factors enhancing the erosion process such as soil type, slope gradient, drainage, geology, and land cover. Multi-temporal satellite images provide valuable information related to seasonal land use dynamics. Satellite data can be utilized for studying erosional features such as gullies, rainfall interception by vegetation, and vegetation cover factor. A Digital Elevation Model (DEM), one of the vital inputs required for soil erosion modeling, is generated by analysis of stereoscopic optical and microwave RS data. RS provides a significant source of real-time and accurate data related to land and soil. It enables similar information over large regions, and can therefore greatly contribute to regional erosion assessment. Multi-temporal satellite images are useful to extract the valuable information associated with seasonal land use dynamics for mapping land use/land cover. It can be used to generate a cover and management factor (C-factor) (Morgan 1995; Wischmeier and Smith 1978) which is one of the input requirements of erosion modeling. The factors associated with soil classification such as soil properties, climate, vegetation, topography, and lithology can be potentially mapped with satellite RS to account for spatial differences in erodibility, which serve as input data for erosion modeling. Optical satellite imagery can be used for soil mapping mainly through the visual delineation of soil patterns (Dwivedi et al. 1997). The basic requirement of any hydrologic or geomorphologic study is a DEM, which allows us to derive various topographic attributes such as elevation, slope, and the aspect which are essential to analyze the physical characteristics of watershed. DEM data is extracted from the satellite images of terrain such as stereo optical imagery (Kumar 2003).

GIS has emerged as a powerful tool for handling spatial and nonspatial geo-referenced data for preparation and visualization of input and output and interaction with soil erosion models. There is considerable potential for the use of GIS technology as an aid to the soil erosion inventory concerning soil erosion modeling and erosion risk assessment. A GIS can be used to scale up to regional levels and to quantify the differences in soil loss estimates produced by different scales of soil mapping used as a data layer in the model. The integrated use of RS and GIS could help to assess quantitative soil loss at various scales and to identify areas that are at potential risk of soil erosion (Saha et al. 1992). Several studies showed the potential utility of GIS techniques for quantitative assessment of soil erosion hazards using various soil erosion models (Saha et al. 1992; Shrestha 1997). Considering the inaccessibility of hilly terrain, if it is an extensive area, RS is essential to accommodate spatial variability and information. Spatial modeling involves the use of GIS to represent the conceptual model and perform simple mathematical computations on the stored GIS object attributes for displaying the results spatially (Kumar and Rastogi 2005).

1.5   Objectives of Study

With the above in mind, the objectives of the present study are:

1. To assess the annual rates of soil erosion from a watershed using distributed information for topography, land use and soil using RS and GIS techniques.
2. To compare the simulated sediment loss with the observed sediment loss.

Chapter 2
USLE–GIS-Based Soil Erosion Assessment: An Overview

Abstract Soil erosion due to human encroachment in the cultivable or forest land is a main concern today due to ever-increasing demographic pressure. Man-made channel and diversion of the natural river flows have resulted in inappropriate drainage that leads to accelerated soil entrainment. Available means to assess the extent of soil erosion in a particular area usually comprises of the combination of the Universal Soil Loss Equation (USLE) and spatial assessment tools. Researchers usually assess soil erosion numerically as well as spatially and validate the output with observed data sets. Rainfall, soil, topography, and ground cover are the key parameters used in distributed soil erosion modelling. Geographic Information System (GIS) software like ArcGIS, ERDAS Imagine, Idrisi, and ILWIS are used to create, store, and analyze the raster layers of USLE factors. Many researchers have contributed to the current state of USLE- and GIS-based distributed soil erosion modelling. Furthermore, many different approaches and techniques have been suggested by researchers to determine USLE factors.

Keywords USLE · GIS · RUSLE · DEM · SDR

Controlling soil loss from watersheds is a challenging problem for watershed managers and soil conservationists. The basic processes and factors that are responsible for inducing land degradation, particularly soil erosion and associated phenomena, are critical to the conceptualization, design, and implementation of productive, stable, and sustainable agricultural systems. Particularly on steep lands, the potential for soil erosion and runoff water losses is high. The productivity and degradation hazards on these lands is determined by the site's climate, soil, and topography. However, their uniqueness lies more with their topographic constraints than with other factors. Use of steep lands is an increasingly common practise in the tropics because of high population pressures and continuing encroachment on hilly lands. Erosion potential and actual erosion in such cases may even exceed hundreds of tons of soil loss per hectare per year. Thus, the selection and design of cropping systems, land management systems, and water management systems must be tailored to attain effective erosion and runoff control to avoid their detrimental impacts

R. J. Patil, *Spatial Techniques for Soil Erosion Estimation*, SpringerBriefs in GIS,
https://doi.org/10.1007/978-3-319-74286-1_2

both on-site and off-site. This chapter briefly describes research work carried out in India and abroad related to the objectives of this study.

2.1 Estimation of Soil Erosion Using RS and GIS Approach

Kothyari and Jain (1997) developed a method for the determination of sediment yield from a catchment using GIS. The method involves spatial disaggregation of the catchment into cells having uniform soil erosion characteristics. The surface erosion from each of the discretized cells was routed to the catchment outlet using the concept of Sediment Delivery Ratio (SDR), which is defined as a function of the area of a cell covered by forest. The sediment yield of the catchment is defined as the sum of the sediments delivered by each of the cells. The spatial discretization of the catchment and the derivation of the physical parameters related to erosion in the cells was performed by a GIS technique using the Integrated Land and Water Information System (ILWIS) package.

Jain and Kothyari (2000) proposed a GIS-based method for the identification of sediment source areas and the prediction of storm sediment yield from catchments. Data from the Nagwa and Karso catchments in Bihar (India) was used. The Integrated Land and Water Information System (ILWIS) GIS package was used for carrying out geographic analysis. An Earth Resources Data Analysis System (ERDAS) image processor was used for the digital analysis of satellite data for deriving the land cover and soil characteristics of the catchments. The catchments were discretized into hydrologically homogeneous grid cells to capture the catchment heterogeneity. The cells thus formed were then differentiated into cells of overland flow regions and cells of flow channel regions based on the magnitude of their flow accumulation areas. The gross soil erosion in each cell was calculated using the Universal Soil Loss Equation (USLE) by carefully determining its various parameters. The concept of SDR was used to determine the total sediment yield of each catchment during isolated storm events.

Fernandez et al. (2002) stated that because of increasing concerns about water quality and aquatic habitat, there is an accute need to quantify and predict sediment yield at a watershed level. Models for sediment yield provide invaluable information when applied to those areas lacking data, for guiding data collection programs and for predicting future impacts of agricultural activities, land use, stream stabilization, and flood control practices. This study was conducted to develop a methodology using GIS and computer modeling to estimate the spatial distribution of soil erosion and sediment yield in the Lawyers Creek Watershed located in western-central Idaho. Soil erosion and the sediment yield for the Lawyers Creek Watershed were estimated by using the Revised Universal Soil Loss Equation (RUSLE) and the Sediment Delivery Distributed (SEDD) model integrated with GIS.

Hoshikawa et al. (2005) carried out an estimation of potential sediment yield by using USLE and GIS integration over eight sub-watersheds: Yokokawa, Katagiri, Matsukawa, Shintoyone, Minowa, Miwa, Koshibu, and Misakubo watersheds of the Tenryu Basin, Japan. The estimated sediment yield was validated with the actual sedimentation measured from their corresponding reservoirs. From the study, it was found that the variation of observed annual sedimentation at reservoirs showed a similar trend with the estimated USLE values within the studied watershed. This outcome advocates the reliability and adaptability of the USLE method that was integrated with GIS for estimating annual potential sediment yield over the large watershed having complex topography and geology.

Lim et al. (2005) stated that accelerated soil erosion is a worldwide problem because of its economic and environmental impacts. To estimate soil erosion efficiently and to establish soil erosion management plans, many computer models have been developed and used. The RUSLE has been used in many countries, and input parameter data for RUSLE have been well established over the years. However, RUSLE cannot be used to estimate the sediment yield for a watershed. Thus, the GIS-based Sediment Assessment Tool for Effective Erosion Control (SATEEC) was developed to estimate soil loss and sediment yield for any location within a watershed using RUSLE and a spatially distributed SDR. SATEEC was enhanced in this study by developing new modules to (1) simulate the effects of sediment retention basins on the receiving water bodies, (2) estimate the sediment yield from a single storm event, and (3) prepare input parameters for the Web-based sediment decision support system using a GIS interface. The enhanced SATEEC system was applied to the study watershed to demonstrate how the enhanced system can be effectively used for soil erosion control. All the procedures were fully automated with the Avenue and database programming; thus, the enhanced SATEEC system does not require experienced GIS.

Kim (2006) used RUSLE and GIS for soil erosion modeling in the Imha watershed, South Korea. The RUSLE model was combined with GIS techniques to analyze the gross soil loss rates caused by typhoon "Maemi" and the annual average soil loss and to evaluate the spatial distribution of soil loss rates under different land uses. The annual average soil loss rate and soil loss rate caused by typhoon Maemi were predicted as 3,450 and 2,920 tons/km^2/year, respectively. Also, the cover management factor for forested areas of the Imha watershed was calibrated using a trial and error method from the relationship between the annual soil losses and various SDR models. The determined C value for the forested area was 0.03 and is three times larger than that of the undisturbed forested area. The SDR was determined to be 25.8% from the average annual rate of soil loss and the surveyed sediment deposits in the Imha.

Htun et al. (2007) carried out spatial pattern analysis of land degradation using Satellite Remote Sensing (SRS) data and GIS over Mandalay watershed, Central Myanmar. This study explored the influence of the main socioeconomic factors on erosion processes and conservation measures in a dry zone farming context to produce an erosion risk map of the study area. It also put forward a way of studying soil erosion and interactions between soil erosion factors by integrating GIS and

RS. Firstly, satellite images (ETM) and the real ground conditions were rectified. Then, according to the ecological and environmental factors, the spatial database and Digital Elevation Model (DEM) of the study area was built. Finally, referring to USLE, the quantization, and distribution of the soil erosion, risks and changes were obtained. The result of this assessment showed the spatial distribution of different land degradation severities across the area on a watershed basis. This study was intended to develop a model based on spatial information and demographic data.

Alejandra (2008) used RS and GIS techniques to estimate soil erosion in the Grande De watershed of Puerto Rico. This work used the USLE to calculate the soil erosion rate. Model inputs such as cover factor and conservation practice factor were successfully derived from remotely sensed data. The LS factor map was generated from slope map. The K factor map was prepared from soil map. Maps covering each parameter (R, K, LS, C, and P) were integrated to generate a composite map of potential erosion intensity based on advanced GIS functionality.

Avanzi et al. (2009) studied soil erosion prediction in the Grande River basin, Brazil using distributed modeling. They stated that mapping and assessment of erosion risk is an important tool for planning natural resources management, allowing researchers to modify land use properly and implement management strategies more sustainably in the long term. This study was particularly aimed at the application of USLE and GIS PC Raster to estimate potential soil loss from the Grande River basin, allowing identification of the susceptible areas to water erosion and estimation of SDR for the adoption of land management practices. For the USLE model, the following factors were used: rainfall–runoff erosivity (R), erodibility (K), topographic (LS), cover management (C), and support practice (P). The Fournier Index was applied to estimate R factor over the basin using six pluviometric stations. Maps of the K, LS, C, and P factors were derived from the soil map, DEM, and land use maps, taking into account information available in the literature. SDR was estimated based on Transported Sediment (TS) to basin outlet and mean soil loss in the basin (MSL) to validate the simulation process. The SDR calculation included data (total solids in the water and respective discharge) between 1996 and 2003 which were measured at a gauging station located on the Grande River and a daily flow data set obtained from the Brazilian National Water Agency (ANA). They concluded that it is possible to validate the erosion process based on the USLE and SDR application for the basin conditions since absolute errors of estimate were low.

Dengiz et al. (2009) employed GIS and RS techniques to assess soil erosion in the Ankara-Guvenc Basin, Turkey. The selected theme layers of this model include a topographic factor, soil factors (depth, texture, impermeable horizon), and land use. Slope layer and land use–land cover data were prepared by using DEM and Landsat-TM satellite imagery. After combining the layers, a soil erosion risk map was produced. The results showed that 44.4% of the study area was at high soil erosion risk, whereas 42% of the study area was insignificantly and only slightly susceptible to erosion risk. It was also found that only 12.6% of the total area was moderately susceptible to erosion risk. Furthermore, conservation land management measures were suggested for average, high, and very high erosion risk areas.

Esther (2009) used GIS techniques to determine RUSLE's "*R*" and "*LS*" factors for the Kapingazi River catchment in Kenya. The purpose of this study was to use GIS techniques to determine some of the soil erosion factors including rainfall–runoff erosivity (R) and slope length/steepness factor (LS). A preliminary assessment of potential soil erosion was carried out to establish the effects of these factors, especially topography. The factors were calculated using the local data that was collected specifically for Kapingazi River Catchment.

Said and Svoray (2009) conducted a multi-criteria analytical study on soil loss, water ponding, and sediment deposition variations using rainfall intensity and land use. They stated that prediction of areas prone to land degradation in agricultural catchments is a complex task due to the difficulties encountered in data gathering over wide regions and in the translation of existing scientific knowledge to a quantitative and spatially explicit risk assessment system. This study incorporated the use of remotely sensed data, terrain analysis, and a multi-criteria mechanism for evaluating risks of soil loss, water ponding, and sediment deposition in a mid-sized agricultural Mediterranean catchment. The research used simulations to study the effect of topographic attributes, soil characteristics, vegetation cover, rainfall intensity, and human activities on the three processes mentioned above. They stated that use of a simple weighted linear combination was more useful than the more sophisticated computerized programming technique. They concluded that an increase in rainfall intensity and land use transformation from orchard to field crops has led to a significant increase in soil loss and sediment yield, while extreme changes in tillage direction have only yielded minor changes in water ponding.

Arekhi et al. (2011) tested the Kinematic Wave-Geomorphologic Instantaneous Unit Hydrograph-Modified Universal Soil Loss Equation (KW-GIUH-MUSLE) model for the estimation of potential sediment yield from the Kengir watershed of Iran. The runoff factor was calculated for six storms in the year 2000. The spatial distribution of the soil erodibility factor was taken from the attribute data about the soils of the study area. The topographic factor (LS) was calculated by multiplying the length (L) and slope (S) factor from the created maps. On the other hand, cropping management and erosion control practice factors for different land use were taken from the satellite-based land use/land cover attribute data and Geographic Information System (GIS), respectively. Sediment yield at the outlet of the study watershed was simulated for six storm events spread over the year 2000 and validated with the measured values. The percent deviations between the sediment yield measurements and observations vary in the range of −22.50 to 5.83%. The high coefficient of determination value (0.99) indicates that the model sediment yield predictions were satisfactory for practical purposes.

Mahmoodabadi (2011) employed the Modified Pacific Southwest Inter-Agency Committee (MPSIAC) semiquantitative model along with GIS and RS techniques to estimate sediment yield in a semiarid region in Central Iran. Nine data layers of the model were generated from Landsat ETM+ imagery, adapted regional maps and field surveys. The GIS was applied to integrate the layers together and generate a sediment yield map. The results showed a range of sediment yield from 263.3 to 496.9 t/km^2/year with an average of 356.4 t/km^2/year. It was stated that this model is less data demanding and provides an efficient way to estimate sediment yield from ungauged basins. It was concluded that hills are the most sensitive land types to sediment yield in the region.

Pal and Samanta (2011) used RS and GIS techniques to estimate soil erosion in the Kaliaghai River basin, West Bengal. The RUSLE model was used for soil loss estimation. Different factors, namely R, K, LS, C, and P, were obtained from monthly and annual rainfall data, soil map, DEM, and RS techniques (with the use of Normalized Difference Vegetation Index) and land use/land cover map respectively.

Amsalu and Mengaw (2014) used the RUSLE model with GIS for soil erosion estimation in Jabi Tehinan Woreda. This study integrated GIS, RS, and Multi-Criteria Evaluation (MCE) techniques to quantify and map erosion vulnerable areas using RUSLE model. Slope gradient, slope length, soil type, soil conservations techniques, cover management, and rainfall variables were used as input model parameters/variables. Finally, the aggregated effects of all parameters were analyzed and soil loss from the area was calculated using RUSEL models. Quantitatively, an estimated annual soil loss in Jabi Tehinan Woreda ranges from nearly 0 in the south and central parts of the area to 504.6 t ha^{-1} year^{-1} in the steep sloping mountainous areas of the north and northeastern parts of the catchment.

Bahrawi et al. (2016) estimated soil erosion using RS and GIS techniques in the Wadi Yalamlam Basin, Saudi Arabia. The RUSLE was used for the assessment of the risk of erosion. Thirty-four soil samples were randomly selected for the calculation of the erodibility factor, based on calculating the K factor values derived from soil properties after interpolating soil sampling points. The soil erosion risk map was reclassified into five erosion risk classes and it was found that 19.3% of the Wadi Yalamlam is under very severe risk (37,740 ha).

Fujaco et al. (2016) developed a GIS script tool based on the USLE to calculate soil loss in three large agricultural sub-basins. Algorithms were implemented in the graphical interface of Model Builder and later in the Python programming language, thus allowing the creation of a specific script to calculate soil loss in an automatic way. The "USLE Paracatu Watershed" script was validated and proved to be effective at estimating erosion in the three sub-basins with an average processing time of half-second per km^2. The friendly interface of the script allows it to be used in any area, only requiring the user to enter the updated data of parameters that compose the equation.

Mali et al. (2016) assessed soil erosion and soil loss risk around hill top surface mines in the Saranda forest, Jharkhand. They employed USLE and the Morgan, Morgan, and Finney (MMF) model to predict the average annual soil erosion over a period of 7 years (2001–2008) using GIS. In the comparative analysis of the 7-year period, the MMF model revealed lower coefficients of variation of 0.71 (2001) and 0.84 (2008) in predicted average annual soil loss, which increased by 16% (81.3–94.2 t ha^{-1} year^{-1}), whereas in the case of the USLE, the coefficients of variation were 3.88 (2001) and 1.94 (2008), with an increase of 61% (48.56–78.38 t ha^{-1} year^{-1}). The correlation coefficients of these models were 0.1 (2001) and 0.36 (2008), which shows that both models predicted significant differences as a result of the different factors considered. Overall, the MMF model predicted a higher soil erosion rate but less variation than USLE.

Uddin et al. (2016) estimated soil erosion dynamics in the Koshi Basin using GIS and RS to assess priority areas for conservation. This study used RS data and a GIS to estimate the spatial distribution of soil erosion across the entire Koshi basin, to identify changes between 1990 and 2010, and to develop a conservation priority map. The RUSLE was used in ArcGIS environment with rainfall erosivity, soil erodibility, slope length and steepness, cover management, and support practice factors as primary parameters. The estimated annual erosion from the basin was around 40 million tons (40 million tons in 1990 and 42 million tons in 2010). The results were within the range of reported levels derived from isolated plot measurements and model estimates. Erosion risk was divided into eight classes from very low to extremely high and mapped to show the spatial pattern of soil erosion risk in the basin between 1990 and 2010. The erosion risk class remained unchanged between 1990 and 2010 in close to 87% of the study area, but increased over 9% of the area and decreased over 3.8%, indicating an overall worsening of the situation. Areas with a high and increasing risk of erosion were identified as priority areas for conservation.

2.2 Estimation of Rainfall Erosivity Factor (R) for USLE

The factor R is the number of rainfall erosion index units (EI_{30}) in a given period at the study location (Subramanya 2009). The rainfall erosion index unit (EI_{30}) of storm is defined as

$$EI_{30} = (KE \times I_{30})/100,$$

where

KE   Kinetic energy of the storm.

The KE in metric tons/ha cm is expressed as

$$KE = 210.3 + 89 \log I,$$

where

I rainfall intensity in cm/h and
I_{30} maximum 30-min rainfall intensity of the storm.

The study period can be a week, month, season, or year. The storm EI_{30} values for that length of the period are summed up.

Annual EI_{30} values are usually computed from data available at various meteorological stations, and similar EI_{30} lines (also known as iso-erodent lines) are drawn for the region covered by the data stations for use in USLE (Subramanya 2008).

Rompaey et al. (2005) conducted a study on modeling sediment yields in Italian catchments. Sediment yield observations derived from 40 long-term sedimentation records in Italian reservoirs were used to calibrate and validate the spatially distributed sediment delivery model WaTEM/SEDEM using the best data available at national scale. Mean annual soil erosion rates for the different catchments were assessed using the USLE-2D procedure. In this model, the rainfall erosivity factor (R) was found to play a vital role and was estimated using following equation:

$$R_{\text{annual}} = \sum_{i=1}^{12} 1.3 P_{\text{monthly}},$$

where

R_{annual} mean annual rainfall erosivity factor (in MJ mm/ha/h/year) and
P_{monthly} the total monthly rainfall (in mm).

Pandey et al. (2007) used USLE and GIS technique for the identification of critical erosion prone areas in the small agricultural watersheds. In this study, the Karso watershed of Hazaribagh, Jharkhand state was divided into 200×200 m grid cells and average annual sediment yields were estimated for each grid cell of the watershed to identify the critical erosion-prone areas of watershed for prioritization purposes. A recent and emerging technology represented by GIS was used as a tool to generate, manipulate, and spatially organize disparate data for sediment yield modeling. USLE was used to predict the spatial distribution of the sediment yield on a grid basis. In this study, the rainfall erosivity factor (R), acts as the key factor of USLE, was computed by using the following formula:

$$\text{KE} = 210.3 + 89 \log_{10} I$$

$$R = \sum \text{Erosion index} = \sum_{i=1}^{n} (\text{KE} \times I_{30}),$$

where

KE   kinetic energy of storm (in MJ/ha),
I_{30} maximum rainfall intensity during a continuous period of 30 min (in mm/h),
I intensity of rainfall (in mm/h) and
R annual erosivity (in MJ mm/ha/h/year).

Dabral et al. (2008) assessed soil erosion in a mountainous catchment of northeastern India using USLE, GIS, and RS. This study aimed to assess the Dikrong River basin of Arunachal Pradesh. The study area was divided into 200×200 m grid cells. ArcInfo 7.2 GIS software and ERDAS IMAGINE 8.4 image processing software was provided with input spatial data and USLE was used to predict the spatial distribution of average annual soil loss on a grid basis. As the watershed has no record of rainfall intensity, monthly rainfall data was used to calculate the rainfall erosivity factor (R) annually using the following relationship developed by Wischmeier and Smith (1978):

$$R = \sum_{i=1}^{12} 1.735 \times 10^{l\left(1.5\ \log_{10}\left(\frac{P_i^2}{P}\right) - 0.08188\right)},$$

where

R rainfall erosivity factor in MJ mm/ha/h/year,
P_i monthly rainfall in mm and
P annual rainfall in mm.

The term $\left(\frac{P_i^2}{P}\right)$ in the above formula is popularly known as the Fournier Index. The above formula is used in annual rainfall erosivity factor (R) estimation when the record of rainfall intensity is not available.

Das (2008) used GIS and RS approaches for the identification of critical erosion-prone areas for watershed prioritization. He stated that the major factors responsible for soil erosion include rainfall, soil type, vegetation, topographic, and morphological characteristics of the basin. Surface erosion and sediment yield quantities were found to have considerable variability due to the spatial variation of rainfall and catchment heterogeneity. The catchment was divided into smaller grid cells of 30×30 m to account for catchment heterogeneity by treating the grid cell as a hydrologically homogeneous area to compute soil erosion and sediment outflow in GIS environment using USLE with Transport Limiting Sediment Delivery (TLSD) concept. In this study, the USLE's rainfall erosivity factor (R) was calculated using the following formulae:
Annual relationship:

$$R = 81.5 + 0.38R_{\mathrm{N}} \quad (340 \le R_{\mathrm{N}} \le 3500 \text{ mm})$$

Seasonal relationship:

$$R = 71.9 + 0.361R_{\mathrm{S}} \quad (293 \le R_{\mathrm{S}} \le 3190 \text{ mm}),$$

where

R average annual/seasonal erosion index,
R_{N} average annual rainfall (mm), and
R_{S} average seasonal rainfall (mm).

2.3   Estimation of Soil Erodibility Factor (*K*)

Phadke and Singh (2006) adopted the USLE and GIS approach for the assessment of soil loss by water erosion in the Jamni River basin situated in the Bundelkhand region of India. They stated that effective control of soil erosion requires the ability to predict the amount of soil loss, which would occur under alternative management strategies and practices. They indicated that soil erodibility factor K is the key input for soil loss prediction models. In this study, the following equation was used for the estimation of K factor:

$$K = 0.01292\big[\big(2.1w^{1.14}\big)(12 - x)\big] + [3.25(y - z) + 2.5(z - 3)],$$

where

x organic matter (%),
w silt (%) = (100 − clay %),
y soil structure code, and
z profile permeability class.

Bienes et al. (2007) studied spatial variability of the soil erodibility parameters and their relationship to the soil map at the subgroup level. This study takes a look at the variation of the parameters related to soil erodibility (fractions of clay, silt, fine and coarse sand; organic matter, permeability, and structure) coming from soil pits from the Community of Madrid's soil map (Spain), according to soil taxonomy at the subgroup level. It draws the conclusion that map erodibility should not be estimated from a soil map because the K factor obtained did not present significant differences among the different types of soil. One or more key factors related with soil erodibility must be taken into account if erodibility maps are to be drawn. This research had shown that silt and structure could be considered as key factors for erodibility maps of the area, but no significant differences have been found in critical factors such as clay or organic matter due to the wide range of data variance. To elaborate erosion risk maps, the use of the K factor from the physiographical map is a good alternative. It was found that erodibility was greater in soils developed over the gypsic material, with a value of 0.63 ± 0.28 than in high plateaus (locally know as alcarrias), with a value of 0.40 ± 0.18. To adequately represent soil erodibility, a Kriging geostatistic technique was used, which reduces the variation of the factors considered when they are found to correlate, as is the case with the parameters deemed to calculate K factor. The following equation (Wischmeier and Smith 1978) was used for the estimation of K factor:

$$100\,K = 10^{-4}\,2.71M^{1.14}(12 - a) + 4.2(b - 2) + 3.23(c - 3),$$

where

K K-factor (t m^2 h ha^{-1} h J^{-1} cm^{-1}),
M texture from the first 15 cm of soil surface
$\quad = [(100 - \text{Ac}) \cdot (L + \text{Armf})],$

where

Ac	% of clay (<0.002 mm),
L	% of silt (0.002–0.05 mm),
Armf	% of very fine sand (0.05–0.1 mm),
a	% of organic matter content,
b	structure of soil (very fine granular: 1–2 mm; fine granular: 2–5 mm; med or coarse granular: 5–10 mm; blocky, platy or massive: >10 mm), and
c	permeability ($c = 1$, very rapid, $c = 2$, mod. to rapid, $c = 3$, moderate, $c = 4$, slow to mod., $c = 5$, slow, $c = 6$, very slow).

Bahrami et al. (2011) developed a nomograph for the estimation of erodibility factor (K) of calcareous soils in Northwest Iran. They stated that for the USLE model, the soil erodibility factor (K) is measured using the average rate of soil loss from the unit plot per unit of rainfall erosivity factor. The USLE nomograph can also estimate this factor based on some measurable soil properties. The USLE nomograph has been developed based on field measurements of soil loss in the semi-humid regions of the USA, where soils were un-calcareous with low values of carbonates (lime). Thus, application of the USLE nomograph in semiarid regions' soils may lead to inaccurate assessment of the K factor. The K factor was measured by natural rainfall events in 36 unit plots from March 2005 to March 2007 and estimated using the USLE nomograph based on soil properties. The results showed that the nomograph-based estimates were 8.77 times more than the measured values. The following multi-regression equation was applied for the estimation of K factor:

$$K = 2.8 \times 10^{-7} M^{1.14}(12 - a) + 4.3 \times 10^{-3}(b - 2) + 3.3 \times 10^{-3}(c - 3),$$

where

K	soil erodibility factor (t h MJ^{-1} mm^{-1}),
M	[(100 − % clay) × (% very fine sand + % silt)],
a	% organic matter,
b	soil structure code, and
c	profile permeability class.

2.4   Estimation of Topographic Factor (LS)

Topographic factor (LS) is the product of the length of slope factor (L) and gradient of slope factor (S). The slope length factor (L) is the ratio of soil loss from the given length of the slope to that from the land having 22.13 m length of the slope if all other conditions remain unchanged. Also, the slope gradient factor (S) is the ratio of soil loss from given gradient of the slope to that from land having 9% slope if all other conditions remain unchanged (Wischmeier and Smith 1978).

Eiumnoh (2000) integrated GIS and SRS for watershed management. The USLE was used for the estimation of soil loss. He stated that the topographic factor (LS) plays a key role in soil detachment down the slope. In this study, DEM was used for the estimation of LS factor. Topographic map was used for elevation data to generate the DEM. Contour lines at 20 m interval in hilly areas and 10 m interval in the plains were digitized from the topographic map using ArcInfo software. The following equation was used for the estimation of LS factor for less than 8% slope gradient:

$$LS = \left(\frac{l}{22.13}\right)^{n}\left(0.0065 + 0.045s + 0.0065s^2\right),$$

where

l slope length (m),
s slope gradient (%), and
n slope length exponent whose maximum value is 0.5 (Morgan 1996).

For greater than 8% slope gradient, the following equation (Liengsakul et al. 1993) was used:

$$LS = \left(\frac{l}{22.13}\right)^{0.5}\left(0.17s - 0.55\right)$$

Gallant et al. (2001) worked on the prediction of the sheet and rill erosion over the Australian continent by incorporating monthly soil loss distribution. They used RUSLE for soil loss estimation. They stated that the topographic factor, (LS) which is the product of slope length factor (L) and slope gradient factor (S), plays a fundamental role in soil erosion by water. It represents the increase in storm runoff volume with increasing hillslope length. For cropping land, L was evaluated by the equation used in RUSLE (Renard et al. 1997) with

$$L = (X_h/22.13)^{m},$$

where

X_h the horizontal slope length (m) and
m variable slope length exponent and m is related to the ratio of rill erosion to inter-rill erosion.

The slope gradient factor (S) was calculated using the following equations:

$$S = 10.8 \times \sin\theta + 0.03 \quad \text{for slope percent} < 9\%$$

$$S = 16.8 \times \sin\theta - 0.50 \quad \text{for slope percent} \geq 9\%,$$

where

θ slope angle (degree).

Batistella et al. (2004) used RS and GIS techniques for soil erosion risk mapping in Rondonia, Brazilian Amazonia, by applying the RUSLE. They stated that the *LS* factor accounts for the effect of topography on erosion in RUSLE. The slope length factor (*L*) represents the effect of slope length on erosion, and the slope steepness factor (*S*) reflects the influence of slope gradient on erosion. DEM was used to generate the topographic factor (*LS*). They used the following equation for calculating *LS* factor (Wischmeier and Smith 1978):

$$LS = \left(\frac{\lambda}{22.13}\right)^{m} \left(65.41 \sin^2 \theta + 4.56 \sin \theta + 0.065\right),$$

where

LS topographic factor,
λ slope length (m),
m an exponent (depends on slope), and
θ slope angle (degree).

Singh et al. (2007) studied soil erosion under simulated conditions for black soils of southern India. They used the USLE for the estimation of soil loss. They stated that the slope is an important factor that enhances the process of erosion. They used the following equation to estimate the slope gradient factor (*S*):

$$S = \frac{0.43 + 0.30s + 0.043s^2}{6.613},$$

where

S slope gradient factor and
s slope (%).

In the next step, they used the modified equation (Wischmeier and Smith 1978) for the estimation of slope gradient factor (*S*) as given below:

$$S = 65.41 \sin^2 \theta + 4.56 \sin \theta + 0.065,$$

where

S slope gradient factor and
θ slope angle (degree).

Douros et al. (2009) used the following relationship (Schmidt et al. 2002) for the calculation of length of slope factor (L) and gradient of slope factor (S):

$$L = 1.4 \left(\frac{As}{22.13} \right)^{0.4}$$

$$S = \left(\frac{\sin \beta}{0.0896} \right)^{1.3},$$

where

A_s specific catchment area (m^2/m) and
β slope angle (degrees).

Bhunia et al. (2012) performed quantitative analysis of relief characteristics of Morobe province of Papua New Guinea. The study analyzed seven topographic parameters, namely absolute relief, relative relief, dissection index, slope, aspect, drainage density, and ruggedness index, for a better understanding of local relief characteristics with some observation of structural landscape pattern. Shuttle Radar Topographic Mission (SRTM) DEM was downloaded from the GLCF website (Global Land Cover Facility, Maryland 2000) which was in Tagged Information File Format (TIFF) with 90-m ground resolution. Raster derived from USGS DEM was converted to a Triangulated Irregular Network (TIN) surface model to remove points from an area of interest from one or more embedded feature classes.

Gelagay and Minale (2016) estimated soil loss using remote and GIS in the Koga watershed, northwestern Ethiopia. The soil loss was estimated by using the RUSLE model. A topographic map of 1:50,000 scale, Advanced Spaceborne Thermal Emission and Reflection Radiometer (ASTER) DEM, digital soil map of 1:250,000 scale, 13 years rainfall records of four stations, and Landsat imagery (TM) with spatial resolution of 30 m were used to derive RUSLE's soil loss variables. The RUSLE parameters were analyzed and integrated using raster calculator in the geo-processing tools in ArcGIS10.1 environment to estimate and map the annual soil loss of the study area. The result revealed that the annual soil loss of the watershed extends from none in the lower and middle part of the watershed to 265 t ha^{-1} year^{-1} in the steeper slope part of the watershed with a mean annual soil loss of 47 t ha^{-1} year^{-1}. The total annual soil loss in the watershed was 255,283 tons, and of these, 181,801 (71%) tons cover about 6691 (24%) ha of land. Most of these soil erosion affected areas are spatially situated in the upper steepest slope part (inlet) of the watershed. These are areas where Nitosols and Alisols with higher soil erodibility character (0.25) values are dominant. Hence, slope gradient and length followed by soil erodibility factors were found to be the main factors of soil erosion.

2.5   Estimation of Crop/Cover Management Factor (*C*)

Corp management factor (*C*) is the ratio of soil loss from given cropping management to that from the land kept continuously fallow if all other conditions remain unchanged (Tamas 2005).

Alejandro and Omasa (2007) estimated the vegetation parameter for modeling soil erosion using Linear Spectral Mixture Analysis of Landsat ETM data. Soil conservation planning often requires estimates of soil erosion at a catchment or regional scale. They stated that predictive models such as USLE and RUSLE are useful tools to generate the quantitative estimates necessary for designing sound conservation measures. However, large-scale parameterization and quantification of factors of the soil erosion model are difficult due to the costs, labor, and time involved. Among the soil erosion parameters, the vegetative cover factor (*C*) has been one of the most difficult to estimate over broad geographic areas. The *C* factor represents the effects of vegetation canopy and ground covers in reducing soil loss. Traditional methods for the extraction of vegetation information from RS data such as classification techniques and vegetation indices were found to be inaccurate. Thus, a new approach based on Spectral Mixture Analysis (SMA) of Landsat ETM data has been developed to map the *C* factor for use in the modeling of soil erosion. A desirable feature of SMA is that it estimates the fractional abundance of ground cover and bare soils simultaneously, which is appropriate for soil erosion analysis. Hence, the *C* factor was estimated by utilizing the results of SMA on a pixel-by-pixel basis.

Hui et al. (2008) developed an approach to compute the *C* factor for USLE using the EOS-MODIS Vegetation Index (VI). The development of the Earth Observing System (EOS) project has resulted in images of high temporal and spatial resolutions. The Moderate Resolution Imaging Spectroradiometer (MODIS) is a key instrument aboard the Terra (EOS AM) and Aqua (EOS PM) satellites, which view the entire earth's surface every 1–2 days, acquiring data in 36 spectral bands. With this, more scientific methods for *C* factor estimation using MODIS VI products by taking account of time variation nature of *C* factor have been developed. The aim of this study was to present a new methodology to compute the *C* factor for soil loss estimation using MODIS level 3 products, which consider the time-varying nature of *C* factor by combining the VI of different periods in a year. Based on EOS-MODIS Vegetation Index (VI), You (1999) proposed another method to determine weightages ofmonthly vgetation indices (those contribute to annual vegetation index) using monthly rainfall and average annualvegetation index.

$$C_{\text{year}} = \left[\left(\sum_{i=\text{AP}}^{\text{JL}} r_i^2/P\right) \times C_1 \left(\sum_{i=\text{A}}^{\text{N}} r_i^2/P\right) \times C_2 + \left(\sum_{i=\text{D}}^{\text{M}} r_i^2/P\right) \times C_3\right] / \left(\sum_{i=\text{J}}^{\text{D}} r_i^2/P\right),$$

where

C_{year}	average annual vegetation cover,
r_i	rainfall amount in each month,
P	annual rainfall amount,
C_1, C_2, and C_3	vegetation cover of April to July, August to November, and December to next March, respectively, and
Ap, Jl, A, N, D, M, and J	short of April, July, August, November, and January.

Karaburun (2010) estimated crop management factor (C) for soil erosion modeling using NDVI in Buyukcekmece watershed. He stated that to take measures in controlling soil erosion, it is required to estimate soil loss over an area of interest. Soil loss due to soil erosion can be estimated using predictive models such as USLE and RUSLE. The accuracy of these models depends on parameters that are used in these equations. One of the most important parameters used in both models is the C factor that represents the effects of vegetation and other land covers. Estimating land cover by interpretation of RS imagery involves the Normalized Difference Vegetation Index (NDVI), an indicator that shows vegetation cover. The C factor values for the Buyukcekmece watershed using NDVI were derived from 2007 Landsat 5 TM Image. The final C factor map was generated using the regression equation in Spatial Analyst tool of ArcGIS 9.3 software. It was found that the north part of watershed had higher C factor values, and that almost 60% of the watershed area had C factor classes between 0.2 and 0.4. The spectral reflectance difference between Near Infrared (NIR) and red was used to calculate NDVI. The formula can be expressed as follows (Jensen, 2000):

$$NDVI = (NIR - Red)/(NIR + Red)$$

The relationship between C and NDVI values was found as

$$C \text{ factor} = \{1.02 - (1.21 \times NDVI)\}$$

Saumer et al. (2010) assessed the USLE's crop and management factor (C) for soil erosion modeling in a large mountainous watershed in central China. They stated that one of the key factors in soil erosion control is the vegetation cover and crop type. However, determining these factors adequately for the use in soil erosion modeling is very time consuming especially for large areas. In this study, the crop and management factor C was calculated using the Fractional Vegetation Cover (CFVC) based on Landsat-TM images for the years 2005, 2006, and 2007. The NDVI was computed for each of the Landsat-TM images based on bands 3 (red band; R) and 4 (near infrared; NIR) using the following equation (Rouse et al. 1974):

$$NDVI = (NIR - R)/(NIR + R),$$

where

NDVI   Normalized Difference Vegetation Index,
NIR   reflectance in Near-Infrared band, and
R   reflectance in the red band.

Then by using the above equation, the *C* factor values were calculated by using the correlation between the NDVI and *C* factor.

Zaeen (2012) employed the RS technique for monitoring the risk of soil degradation using NDVI. He stated that one of the most important parameters in soil erosion prediction models is the *C* factor that represents the effects of vegetation and other land covers. In this study, the crop management factor *C* was estimated with the help of NDVI, which shows the effect of vegetation cover. The NDVI of the study area was derived from the Landsat ETM images of the study period. Finally, the *C* factor map of the area was prepared in spatial analyst toolbox of ArcGIS using the following exponential equation:

$$C = e^{\left(-\alpha \frac{\text{NDVI}}{\beta - \text{NDVI}}\right)},$$

where
α and β are the parameters determining the shape of NDVI-*C* curve.

Gunawan et al. (2013) carried out RS and GIS-based soil erosion estimation for supporting integrated water resources conservation and management in the Manjunto watershed, Bengkulu Province, Indonesia. The aim of this research was to assess the average annual rate of potential soil erosion in Manjunto watershed for each soil mapping unit using RS data, namely Normalized Difference Vegetation Index (NDVI) and slope. The NDVI value was obtained from satellite image processing while the slope value was obtained from Shuttle Radar Topographic Mission (SRTM) DEM processing. The results showed that the eroded catchment area increased significantly. The average annual rate of potential soil erosion in Manjunto watershed in the year 2000 amounted to $3 \text{ t ha}^{-1} \text{ year}^{-1}$, while in the year 2009, there was a significant increase to $27.03 \text{ t ha}^{-1} \text{ year}^{-1}$.

2.6 Estimation of Conservation/Support Practice Factor (*P*)

The conservation or support practice factor (*P*) is the ratio of soil loss from a land having specified conservation practices to that from a land plowed in a direction parallel to a slope if all other conditions remain unchanged (Wischmeier and Smith 1978).

Roslan and Tew (1997) used satellite imagery to determine the land use management factor of the USLE. The study was conducted over Cameron Highlands in Malaysia as it is being threatened by landslips and flash floods due to numerous development projects and intensive agricultural activities. A study of the land use management factor [cover and management factor (C) and support practice factor (P)] of the USLE has been undertaken, as this parameter reflects the land cover in the study area and its effect on soil erosion. Using RS satellite imagery, the Ringlet area in the Cameron Highlands was identified as having the highest erosion risk based on the C and P factors, and its highest ranking confirmed this for residential and construction areas compared to other locations. The results of this study highlight the important land uses associated with erosion risk and indicate that development was carried out to ensure a quality environment for the future. They stated that with recent developments in RS technology, color infrared imagery could be used to determine the combined land use management factor (C and P) of the USLE.

Gallant et al. (2001) worked on the prediction of the sheet and rill erosion over the Australian continent by incorporating monthly soil loss distribution. They stated that the supporting practice factor (P) accounts for the effects of contour, strip cropping, or terracing. It is defined as the ratio of soil loss with a specific supporting practice to the corresponding soil loss with certain cultivation. Due to a lack of spatial data on existing contour locations and tillage practices, it was assumed that the values of P factor were one everywhere. Given this scenario, the estimated soil loss rate reflects erosion potential under current conditions with no soil conservation support practices.

Desilva et al. (2010) worked on water erosion modeling in a watershed under forest cultivation through the USLE model. They stated that modeling of erosion processes integrated with a GIS is an important tool to assess erosion by water. The objective of this study was to determine the spatial distribution of water erosion in forest ecosystems and generate soil loss prediction maps according to different land use scenarios. The study was conducted in a watershed occupied by Eucalyptus cultivation located in Belo Oriente in the Rio Doce River valley, Central-East region of Minas Gerais state, Brazil. For the spatial distribution modeling of soil loss for the watershed, the USLE model was coupled with GIS. The factor P, which represents the effect of different land uses, was assumed to be 0.5 for the scenario with conservationist Eucalyptus and 1.0 for the other scenarios.

Sheikh et al. (2011) integrated GIS and USLE for soil loss estimation in a Himalayan watershed. They stated that erosion calculation requires an enormous amount of information and data, usually coming from different sources in various formats and scales. Therefore, GIS was used, which helped considerably in organizing the spatial data representing the effects of each factor affecting soil erosion. The factors that most influence soil erosion are linked to topography, vegetation type, soil properties, and land use/cover. The support practice factor indicates the rate of soil loss according to the various cultivated lands on the earth. P values range from 0 to 1, whereby the value 0 represents an excellent man-made erosion resistance facility and the value one indicates no man-made erosion resistance facility. In the study area, there were some agricultural support practices such as contour farmland and terraced farmland. However, P factor value was taken as 1 for simplicity in this study.

Arekhi et al. (2012) used RUSLE, RS, and GIS for mapping soil erosion and sediment yield susceptibility in Cham Gardalan watershed, Iran. They stated that the P factor reflects the impact of support practices dealing with the average annual erosion rate. The lower the *P* value, the more efficient the conservation practice is deemed to be at reducing soil erosion. As with the other factors, the *P* factor differentiates between cropland and rangeland or permanent pasture. This study considered that there were no conservation practices (*P*) adopted in the area. Therefore, the maximum value for *P,* i.e. 1, was assigned to areas with no conservation practice.

2.7 RS Products for Soil Erosion Estimation

RS plays a unique role in the generation of input data sets for the spatially distributed soil erosion modelling. Many RS products available today are being widely used for soil erosion modelling. A large number of earth observation satellites provide images of the earth surface. These images include information about oceans, vegetation, land use and cover, elevation, biomass, cloud cover, water bodies, atmospheric temperature, etc. A part of this information is potentially useful for distributed soil erosion modelling. This section describes commonly used RS products for soil erosion estimation.

Commonly used satellite products are listed in Table 2.1. These products involve satellites and onboard sensors. The satellite products shown in Table 2.1 include different sensors like Multispectral Scanner (MSS), Thematic Mapper (TM), Enhanced Thematic Mapper plus (ETM+), Operational Land Imager (OLI), Thermal InfraRed Sensor (TIRS), High Resolution Visible (HRV), High Resolution Visible InfraRed (HRVIR), Linear Imaging Self-Scanner (LISS), Advanced Spaceborne Thermal Emission and Reflection Radiometer (ASTER), Advanced Very High Resolution Radiometer (AVHRR), Panchromatic (PAN), and Multispectral sensors. These sensors are generally divided in two categories, optical sensors and imaging radar, based on their sensing characteristics. Optical satellite systems are most commonly used in erosion research (Vrieling 2007). These sensors cover the parts of the electromagnetic spectrum including the Visible and Near Infrared (VNIR) ranging from 0.4 to 1.3 μm, the Shortwave Infrared (SWIR) between 1.3 and 3.0 μm, and the Thermal Infrared (TIR) from 3.0 to 15.0 μm (Vrieling 2007).

Among all the RS satellites, Landsat is the widest used satellite, as it has the longest time series of data among currently available satellite products. Demand for high resolution satellite imageries has been steadily increasing. IKONOS and QuickBird are the two highest resolution satellites available and their products are being used for soil erosion modelling. Many countries are planning to launch a new series of RS satellites to fulfill increasing demands of high resolution (spatial and temporal) data products and to replace current orbiting satellites. Some examples of future satellites include, Landsat 9, which is planned to be launched by 2023 and

Table 2.1 Details of commonly used RS satellite products for soil erosion research

Satellite	Operation period	Sensor	Spatial resolution	Swath width (km)	No of spectral bands	Spectral domain
Landsat-1, 2, 3	1972–1983	MSS	80 m	185	4	VNIR
Landsat-4, 5	1982–1999	TM	30 m	185	6	VNIR, SWIR
			120 m		1	TIR
Landsat-7	1999–present	ETM+	15 m	183	1	VNIR
			30 m		6	VNIR, SWIR
			60 m		1	TIR
Landsat 8	2013–present	OLI, TIRS	15 m	185	1	SWIR
			30 m		9	SWIR
			100 m		2	LWIR
SPOT-1, 2, 3	1986–2009	HRV	10 m	60	1	VNIR
			20 m		3	VNIR
SPOT-4	1998–2013	HRVIR	10 m	60	1	VIS
			20 m		4	VNIR, SWIR
IRS-1A, 1B	1988–2003	LISS-1	72.5 m	148	4	VNIR
		LISS-2	36.25 m		4	VNIR
IRS-1C, 1D	1995–2010	PAN	5.8 m	70	1	VNIR
		LISS-3	23.5 m	142	3	VNIR
			70 m		1	SWIR
IRS-P6 (Resourcesat-1)	2003–2013	LISS-3	23.5 m	140	4	SWIR
Terra	1999–present	ASTER	15 m	60	3	VNIR
			30 m		6	SWIR
			90 m		5	TIR
NOAA/TIROS	1978–present	AVHRR	1.1 km	833	5	VNIR, SWIR, TIR
IKONOS	1999–2015	Panchromatic	1 m	11.3	1	VNIR
		Multispectral	4 m		4	VNIR
QuickBird	2001–2014	Panchromatic	0.61 m	18	1	VNIR
		Multispectral	2.44 m		4	VNIR

Resourcesat-3 of the Indian space Research Organization (ISRO) which is slated to be launched by 2021 to strengthen the fleet of the Indian Remote Sensing Satellites (IRS).

From the review mentioned above, it can be inferred that predictive models like the USLE are most widely used to predict erosion rates in an ungauged watershed. In fact it is universally accepted that the USLE can also form the basis of soil erosion modeling. Several researchers reported that the rate of soil erosion from a watershed depends on rainfall, soil, topography, cropping patterns, and vegetation cover. Some of these factors are also responsible for sediment detachment and transportation processes. However, a collection of data regarding soil, topography, cropping patterns, and ground cover by traditional methods is very time consuming. With the advent of RS techniques, the data collection process has become much more convenient and quick. Some researchers have reported that collection of data for distributed features like soil, topography, cropping pattern, and ground cover has become easier due to the integration of RS and GIS. Nowadays, researchers are widely using RS and GIS techniques along with predictive model like USLE for soil erosion estimation in the watershed.

Chapter 3
Study Area and Data Collection

Abstract The present study was conducted in the Shakkar River watershed. The Shakkar River is a major tributary of the main Narmada River. The study area contains dense forest cover and steeply sloping hillocks formed of marble stone. Human-induced activities, such as urbanization, encroachment into the agricultural area, and stone quarrying/mining, are responsible for removing the protective soil cover (forest/vegetation) and have resulted in a severe state of erosion in the area over the past few decades. Daily rainfall data, soil maps, a digital elevation model, satellite imagery of the study area and observed sediment data were used in the present analysis.

Keywords Shakkar River · Narmada River · Human encroachment
Soil map · Satellite image

The correctness of any prediction tool depends largely upon the accuracy of data sets with relevant information and the methodology adopted. In modeling hydrologic processes such as soil erosion and causative factors such as rainfall, soil type, topography, cropping pattern, and vegetation or ground cover are to be determined along with the hydrologic time series data. However adequate data for a complete and comprehensive analysis is seldom available, leaving no other choice than to use only the available information. This chapter deals with the data collection for the estimation of the annual rate of soil erosion of the Shakkar River watershed lying in the Narmada River basin situated in the Narsinghpur and Chhindwara districts of Madhya Pradesh (India).

3.1 Study Area and Data Availability

3.1.1 About Study Area

The Shakkar River rises in the Satpura range, east of the Chhindi village, Chhindwara district, Madhya Pradesh. The area lies between 22° 20′N to 23° 00′N latitudes and

R. J. Patil, *Spatial Techniques for Soil Erosion Estimation*, SpringerBriefs in GIS,
https://doi.org/10.1007/978-3-319-74286-1_3

78° 40′E to 79° 20′E longitudes with an elevation ranging from 314 to 1154 m above MSL (mean sea level). The watershed covers 2223 km^2 of the total geographical area up to the gauging point. Tectonically, the area belongs to the Central Indian Tectonic Zone (CITZ) of the Satpura Mobile Belt. The Shakkar River, after covering first reaches on basalts, takes a turn from its westerly direction to north and cuts across the Satpura Mountain south of the Narmada River. Moving north and then northwesterly, it comes in the openness of the Narmada alluviums leaving the mountain in the backdrop. The alluvial plain of Narmada is of accretion type, and the Shakkar deposits much of its load here before joining the master river Narmada. The Narmada River, in fact, flows through a famous lineament known as the Son-Narmada lineament. The course of the Narmada is bounded by the Son-Narmada North Fault (SNNF) and Son-Narmada South Fault (SNSF) of the CITZ. The climate of the basin is dry except for the southwest monsoon season. During the southwest monsoon season, the relative humidity exceeds 87% (August month) although the rest of the year is drier. The driest part of the year is the summer season when the relative humidity is less than 33%. May is the driest month of the year. The normal maximum temperature during the month of May is 42.5 °C, and the minimum during the month of January is 8.2 °C. Soils are mainly clayey to loamy in texture with calcareous concretions invariably present (Patil et al. 2015). They are sticky and, due to shrinkage in summer, develop deep cracks. They predominate in montmorillonite and beidellite type of clays. In the other alluvial areas, mixed clay, black to brown and reddish brown soil, derived from sandstones and traps is observed which is sandy clay in nature with calcareous concretions. Near the banks of the river and at the confluence, light yellow to yellowish brown soils are noticed which were deposited in the recent past. These soils are clayey to silty in nature. The area receives rainfall via the southwest monsoon. The southwest monsoon starts from the middle of June and lasts until the end of September. October and middle of November constitute the post monsoon or retreating monsoon season. The average annual rainfall of the area is 1245 mm, whereas normal annual rainfall of the area is 1192.1 mm (Patil et al. 2015). The location of the study area is shown in Fig. 3.1.

3.1.2   Data Availability

This study requires data regarding rainfall, topography, soil, and vegetation or ground cover of the area. The data of observed sediment loss at the outlet of the watershed is also required for validation of the estimation model.

3.1.2.1   Rainfall Data

The rainfall data is needed to estimate the value of rainfall erosivity of the area. Three major rain gauge stations, namely Gadarwara and Harrai falling in the study area and Amarwara in the vicinity of the area, were selected. The 15-year 24-h

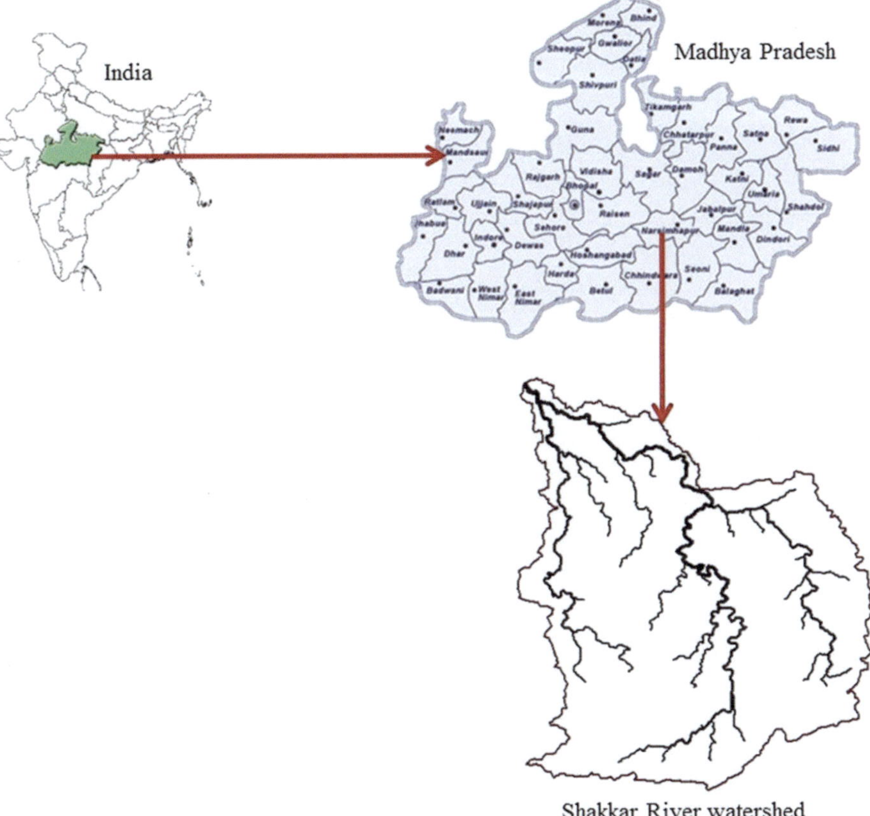

Fig. 3.1 Location map of the study area

rainfall data of these three stations was collected from the Department of Physics and Agrometeorology, College of Agricultural Engineering, Jawaharlal Nehru Agricultural University, Jabalpur (MP). The annual rainfall of these stations was obtained by summing the daily rainfall data, which is presented in Table 3.1.

3.1.2.2   Digital Elevation Model (DEM)

For delineation of the Shakkar River watershed and the preparation of the drainage map, information regarding the topography was needed. In this study, a geo-coded Digital Elevation Model (DEM) generated from Advanced Space-borne Thermal Emission and Reflection Radiometer (ASTER) data was used. The digital elevation model (DEM) was downloaded from the GLCF website (Global Land Cover Facility, Maryland 2000) which was in Tagged Information File Format (TIFF) with 30 m ground resolution.

Table 3.1 Annual rainfall (mm) at respective rain gauge stations

Year	Station name		
	Gadarwara	Harrai	Amarwara
1992	909	1015	844
1993	1279	1276	1258
1994	860	1794	1532
1995	733	951	1085
1996	558	834	899
1997	951	1436	1907
1998	395	1011	1212
1999	2210	2023	1713
2000	928	1018	797
2001	993	1149	625
2002	1031	965	961
2003	1033	1357	938
2004	889	973	601
2005	1279	1202	1392
2006	684	1101	1351
Average	913.26	1207	1141

3.1.2.3   Soil

In this study, soil data was used to estimate the susceptibility of the soils of the study area to erosion by rainfall. For this purpose, the soil map of Madhya Pradesh (sheet no. 5) was used. This map was generated by the National Bureau of Soil Survey and Land Use Planning (NBSSLUP) using the Soil Atlas of Madhya Pradesh. There are nine sheets in total in which the soils of Madhya Pradesh are covered. This map is available at the scale of 1:500000.

3.1.2.4   Satellite Data

To obtain information about the vegetation or ground cover, the land use/land cover map of the study area was prepared with the help of ERDAS IMAGINE 9.2, which is well-known image-processing software. For the preparation of the land use/land cover map, a satellite image of the study area was used, which was obtained from the National Remote Sensing Agency (NRSA, Hyderabad). The details of the satellite data are shown in Table 3.2. False color composite (FCC) of the study area is shown in Fig. 3.2.

Table 3.2 Details of satellite data

Satellite	Sensor	Spatial resolution	Bands	Swath	Row/path	Acquired on
IRS-P6	LISS III	23.5 m	2, 3, 4, 5	141 km	99/56	Jan 8, 2011

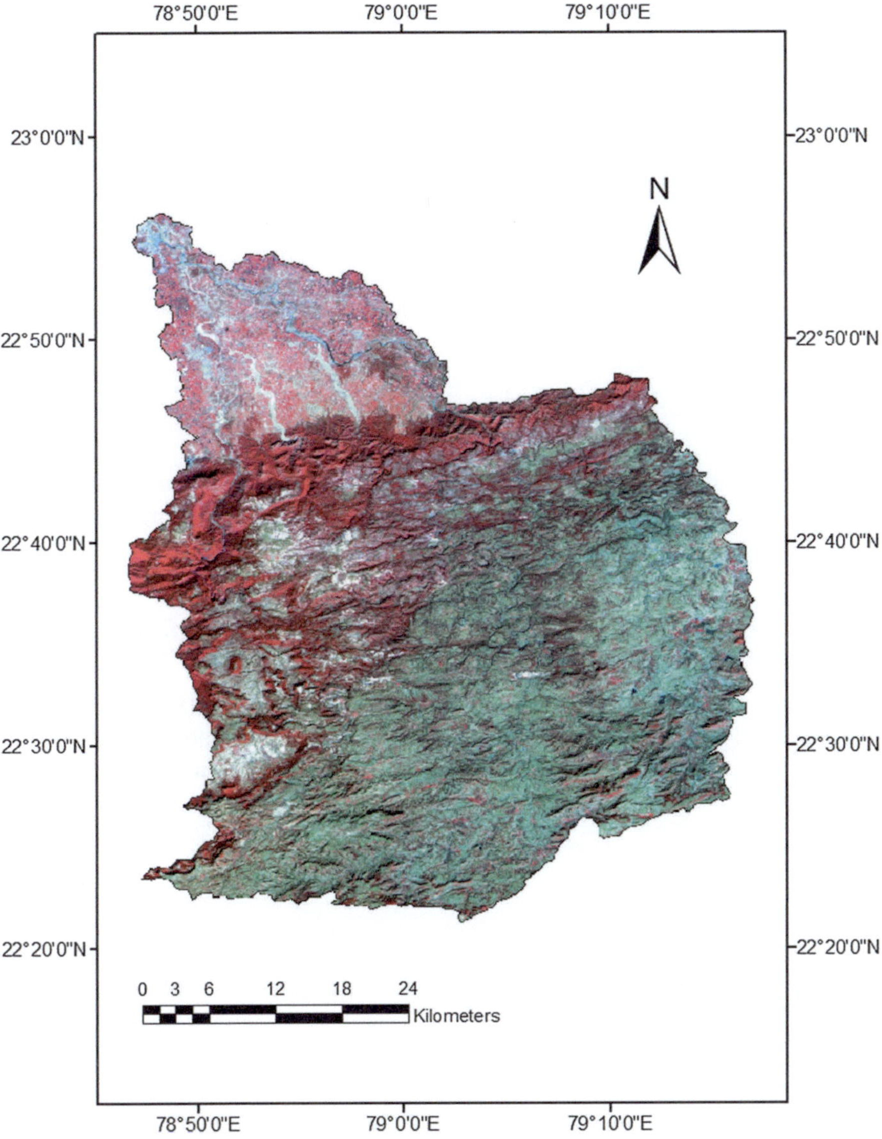

Fig. 3.2 False color composite (FCC) of the study area

3.1.2.5   Observed Sediment Data

Observed sediment data of the stream gauging station at the outlet of the watershed
was collected from the "Integrated Hydrological Data Book for non-classified river
basins" published by Central Water Commission (CWC), New Delhi. The data

Table 3.3 Software used and their distinctive features

Software	Description
ArcGIS 9.3	ArcGIS is one of the most complete and extensible GIS toolkits available. It is highly functional and contains advanced geoprocessing and data conversion capabilities. Professional GIS users use ArcMap (ArcInfo) for all aspects of data building, modeling, analysis, and map display for the screen and output. ArcCatalog is mostly used for creating, deleting, and editing the spatial data files (ESRI).
ERDAS IMAGINE 9.2	The ERDAS IMAGINE software performs the functions of both image processing and geographic information systems (GIS). These functions include importing, viewing, altering, and analyzing raster and vector data sets.
Other software	Window-based software such as MS office was used to build a database and analyze the data.

books published in the years 2006 and 2012 were used. These data books contain statistics regarding annual sediment loss at the outlet (at the gauging point).

3.1.3 Software Used for Preparation of Thematic Maps

Various software packages available at the well-established GIS lab of Madhya Pradesh Water Sector Restructuring Project (MPWSRP), Department of Soil and Water Engineering, College of Agricultural Engineering, Jabalpur were used for the preparation of thematic maps and analysis of data. Details of the software packages used are given in Table 3.3.

Chapter 4
Spatial Techniques for Soil Erosion Estimation

Abstract Soil is an essential resource for human livelihoods. Soil erosion is now a global environmental crisis that threatens the natural environment and agriculture. This study aims to assess the annual rates of soil erosion using distributed information for topography, land use, and soil, with a Remote Sensing (RS) and Geographical Information System (GIS) approach and comparison of simulated sediment loss with observed sediment loss. The Universal Soil Loss Equation (USLE) with RS and GIS techniques was used to predict the spatial distribution of soil erosion occurring in the study area on a grid-cell basis. Thematic maps of rainfall erosivity factor (R), soil erodibility factor (K), topographic factor (LS), crop/cover management factor (C), and conservation/support practice factor (P) were prepared using annual rainfall data, soil map, Digital Elevation Model (DEM) and an executable C++ program, and a satellite image of the study area in the GIS environment. This study puts an emphasis on the use of a C++ program to estimate the topographic factor (LS).

Keywords Soil erosion · USLE · Remote Sensing · GIS · Digital Elevation Model (DEM) · C++ program

4.1 Introduction to Geographic Information Systems (GIS)

Development of Geographic Information Systems (GIS) closely follows advancements in computers. As computers can handle more data-intensive operations, the use of GIS has also expanded to handle larger data sets. GIS are primarily used to process and display data, which have a spatial component. The spatial information determines where the data model is located in the real world. The object's attributes

---

The original version of this chapter was revised: Copyright text "© IAHS Press" has been added in the figure caption. The erratum to this chapter is available at https://doi.org/10.1007/978-3-319-74286-1_7

Fig. 4.1 Example of GIS
data layers organized into
separate themes

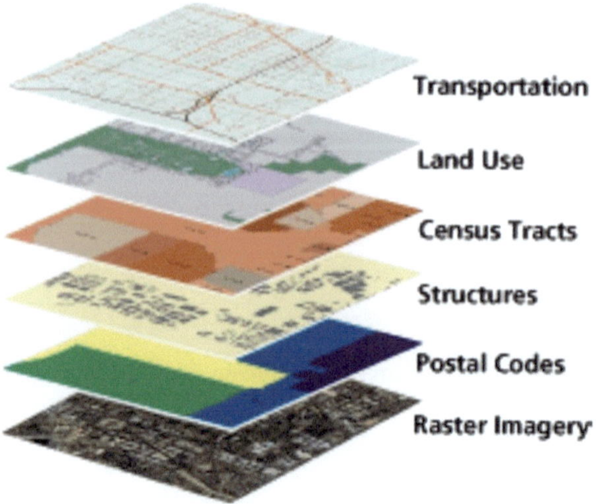

or specific characteristics are also contained within the data model. Attributes such as length, area, and count are important to distinguish between data models. Current GIS software packages are capable of storing complex spatial information into separate, thematic layers (Fig. 4.1).

The two spatial data types used in this research project are vector and raster files. Vector data contain features defined by a point, line, or polygon. Vector data models are useful for storing and representing discrete features such as buildings and roads. ArcGIS implements vector data as shapefiles. Raster data are composed of a rectangular matrix of cells. Each cell has a width and length, and is a portion of the entire area represented by the raster. Each cell has a value, which represents the phenomenon portrayed by the raster data set, such as a category, magnitude, distance, or spectral value. The category could refer to a land use class, such as grassland or urban. The cell size dimensions can be as large or as small as necessary, to accurately represent the area. The location of each cell is defined by either its reference system or projection. The use of the same projection system allows one raster layer to overlap another layer. This research has used the Datum Indian 1975 (D_Indian_1975) and Indian Polyconic projection system for all data types.

4.2 Methods of Analysis

For accurate predictions, the relevant data and methods of analysis are two essential tools. This section includes the theory and procedure used to estimate the annual rates of soil erosion using the Universal Soil Loss Equation (USLE) with remote sensing and GIS approaches. The detailed procedure for estimating various USLE factors is given in the following subsections.

4.2.1 USLE-Based Spatially Distributed Soil Erosion Estimation

The rate of soil erosion from an area is strongly dependent upon rainfall, topographic characteristics, soil, and vegetation. These factors vary significantly within a watershed. Therefore, a method, which takes these factors into account while estimating soil erosion, is expected to produce realistic estimates of rates of soil erosion (Das 2008). The USLE (Wischmeier and Smith 1965) is one such equation that takes factors such as rainfall, topography, soil, and land use into consideration while assessing soil erosion. The USLE is a simple empirical model used extensively to realistically estimate surface soil erosion over small areas (Jain and Kothyari 2000; Yusof and Baban 2002; Jain and Goel 2002; Mishra et al. 2007).

4.2.2 The USLE

The USLE (Wischmeier and Smith 1965) for estimating the average annual rate of soil erosion can be expressed as

$$A_i = R_i.K_i.L_i.S_i.C_i.P_i, \tag{4.1}$$

where

A_i average annual rate of soil loss (t/ha/year),
R_i rainfall erosivity factor (MJ mm ha^{-1} h^{-1}),
K_i soil erodibility factor (t ha h ha^{-1} MJ^{-1} mm^{-1}),
L_i slope length factor (Dimensionless),
S_i slope steepness/slope gradient factor (Dimensionless),
C_i crop/cover management factor (Dimensionless), and
P_i conservation/support practice factor (Dimensionless) of the ith cell.

In a large-sized watershed, these factors (R, K, LS, C, and P) show spatial variability. Thus, a watershed needs to be discretized into smaller homogeneous areas to capture catchment heterogeneity (Jain and Kothyari 2000). Several methods are available for discretization of the watershed into smaller areas. The cell or grid approach is mostly used due to its adaptability to raster-based GIS and practicality in the collection of input data using remotely sensed satellite data. A cell size of 30 m × 30 m is widely used in distributed modeling (Kothyari et al. 2002; Fernandez et al. 2003; Staples et al. 2012; Zivotic et al. 2012; Patil et al. 2015). The same cell size is used to represent the cell as a hydrologically homogeneous area in the present study as well. Grids thus formed can be categorized as lying on overland areas and those lying in channel areas. Such a differentiation is necessary because the processes of sediment loss and delivery are widely different (Alejandra 2008). The detailed procedure for estimating USLE factors and preparing thematic maps of these factors is given below.

4.3 Development of Model Database for USLE

4.3.1 Rainfall Erosivity Factor (R)

Erosion by fluvial processes is believed to be the single most important factor shaping the earth's subaerial landscape. The general factors influencing erosion are well known, but erosive properties of rain play a vital role in soil detachment by impact action (Phillips 1990).

The Rainfall erosivity (R factor) is one of the most basic and important factors in the assessment of soil erosion using a mathematical model like USLE. Erosivity is the potential capacity of the raindrops to cause detachment of the soil particles from its location, which depends on rainfall intensity and its recurrence. Hence accurate computation of erosivity is necessary for realistic and quantitative estimation of soil erosion. The R factor is defined as the mean annual sum of individual storm erosion index values, EI_{30}, where E is the total storm kinetic energy and I_{30} is the maximum rainfall intensity in 30 min. The continuous rainfall intensity data are needed to compute EI_{30} values of the storm (Elangovan and Seetharaman 2011).

Wischmeier and Smith (1978) state that the factor R is the number of rainfall erosion index units (EI_{30}) in a given period at the study location. The rainfall erosion index unit (EI_{30}) of a storm is defined as

$$EI_{30} = (\text{KE} \times I_{30})/100, \tag{4.2}$$

where

KE   kinetic energy of the storm.

The KE in metric tons/ha cm is expressed as

$$\text{KE} = 210.3 + 89 \ \log I, \tag{4.3}$$

where

I rainfall intensity in cm/h and
I_{30} maximum 30 min rainfall intensity of the storm.

The study period can be a week, month, season, or a year. The storm EI_{30} values for that length of the period are summed up.

Annual EI_{30} values are usually computed from data available at various meteorological stations, and lines of equal EI_{30} lines (also known as iso-erodent lines) are drawn for the region covered by the data stations for use in USLE (Das 2010).

Realistic estimation of monthly or annual rainfall erosivity (EI_{30}) value requires long-term pluviographic data at 15 min interval or less (Wischmeier and Smith 1978). In many parts of the world, especially in developing countries, the spatial coverage of pluviographic data is often difficult to obtain (Yu et al. 2001; Cohen et al. 2005). Monthly, seasonal, and annual rainfall data are usually available for extended periods of time which are used to calculate R factor (Mati et al. 2000; Babu et al. 2004). In India, Babu et al. (2004) computed EI_{30} values using the storm data (greater than 12.5 mm) of 123 rain gauge stations located in different zones of India using Eqs. (4.2) and (4.3). Linear relationships were established between average annual and seasonal (June–September) rainfall with computed EI_{30} values for different zones of India. Derived relationships are as follows:

Annual relationship:

$$R = 81.5 + 0.38R_{N} \quad (340 \leq R_{N} \leq 3500 \text{ mm}) \tag{4.4}$$

Seasonal relationship:

$$R = 71.9 + 0.36R_{S} \quad (293 \leq R_{S} \leq 3190 \text{ mm}), \tag{4.5}$$

where

R average annual/seasonal erosion index,
R_{N} average annual rainfall (mm), and
R_{S} average seasonal rainfall (mm).

These relationships are valid within the indicated rainfall range (Eqs. 4.4 and 4.5). The above-derived regression equations and average annual/seasonal (June–September) rainfall data were used to approximate the erosion index values of 500 locations evenly distributed all over India, and iso-erodent maps were drawn for annual and seasonal EI_{30} values (Babu et al. 2004). For the present study, Eq. (4.4) was used to compute annual values of the rainfall erosivity (R) factor by replacing R_{N} with actual rainfall in a year in the watershed.

Finally, a thematic map of rainfall erosivity factor (R) was prepared in the GIS environment. For this, a Thiessen polygon of the study area was developed using the spatial analyst toolbox of ArcGIS 9.3. Three rain gauge stations, namely Gadarwara, Harrai, and Amarwara, were used for the preparation of the Thiessen polygon. The annual R factor values at individual stations were estimated using the above Eq. (4.4). After attributing these values to the Thiessen polygon, raster maps of the R factor for individual years were prepared using the conversion toolbox of ArcGIS 9.3.

4.3.2   *Soil Erodibility Factor* (**K**)

The soil erodibility factor (K) is the rate of soil loss per rainfall erosion index unit as measured on a unit plot (Wischmeier and Smith 1978). The soil erodibility factor

(K) represents the effect of soil properties and soil profile characteristics on soil loss. Soil properties are one of the main parameters that affect runoff and soil erosion processes (Kavian et al. 2014). In practical terms, the soil erodibility factor is the average long-term soil and soil profile response to the erosive powers of rainstorms; i.e., the soil erodibility factor is a lumped parameter that represents an integrated average annual value of the total soil and soil profile reaction to a significant number of erosion and hydrologic processes. These processes consist of soil detachment and transport by raindrop impact and surface flow, localized deposition due to topography and tillage-induced roughness, and rainwater infiltration into the soil profile (Young et al. 1987).

Wischmeier and Smith (1978) suggested the following equation for the estimation of the soil erodibility factor (K):

$$100\,K = 10^{-4}\,2.7 \cdot 1M^{1.14}(12 - a) + 4.2(b - 2) + 3.23(c - 3), \qquad (4.6)$$

where

K K-factor (t m^2 h ha^{-1} hJ^{-1} cm^{-1}),
M texture from the first 15 cm of soil surface

$$= [(100 - \mathrm{Ac}) \cdot (L + \mathrm{Armf})],$$

where

Ac   % of clay (<0.002 mm),
L % of silt (0.002–0.05 mm),
Armf   % of very fine sand (0.05–0.1 mm),
a % of organic matter content,
b structure of soil (very fine granular: 1–2 mm; fine granular: 2–5 mm; med or coarse granular: 5–10 mm; and blocky, platy, or massive: >10 mm), and
c permeability ($c = 1$, very rapid; $c = 2$, moderate to rapid; $c = 3$, moderate; $c = 4$, slow to moderate; $c = 5$, slow; and $c = 6$, very slow).

The K factor values are usually determined at experimental runoff plots or using empirical erodibility equations, which relate several soil properties to the K factor (Subramanya 2009). Direct determination of the K factor requires long-term measurements of soil loss, which is costly and time-consuming, and shows very poor correlation with values determined by empirical erodibility equations (Vaezi et al. 2011). Thus, in this study, the K factor values were selected from literature values (Fistikoglu and Harmancioglu 2002) provided by Singh et al. (1992). The soil map of Madhya Pradesh (at 1:500000 scale) prepared by the National Bureau of Soil Survey and Land Use Planning (NBSSLUP 1996) was used to obtain soil textural and related information of the soil map units of the study area. The individual soil map units were digitized in the ArcMap extension of ArcGIS 9.3. Then, the respective values of the K factor for each soil map unit were attributed to the digitized map and a raster map of the K factor was prepared using the conversion toolbox of ArcGIS 9.3.

4.3.3 Topographic Factor **(LS)**

Topographic factor (LS) is the product of slope length factor (L) and slope steepness/slope gradient factor (S).

4.3.3.1 **Slope Length Factor** *(L)*

Slope length can be defined as the distance from the point of origin of overland flow to the point where either the slope gradient decreases enough that deposition begins, or the runoff water enters a well-defined channel (Wischmeier and Smith 1958) as shown in Fig. 4.2.

Slope length factor (L) is the ratio of soil loss from a given length of slope to that from land having 22.13 m length of slope. It is expressed as (Wischmeier and Smith 1978)

$$L = \left(\frac{\lambda}{22.13}\right)^{m},\tag{4.7}$$

where

λ field slope length in meters and
m an exponent (value ranging from 0.2 to 0.5).

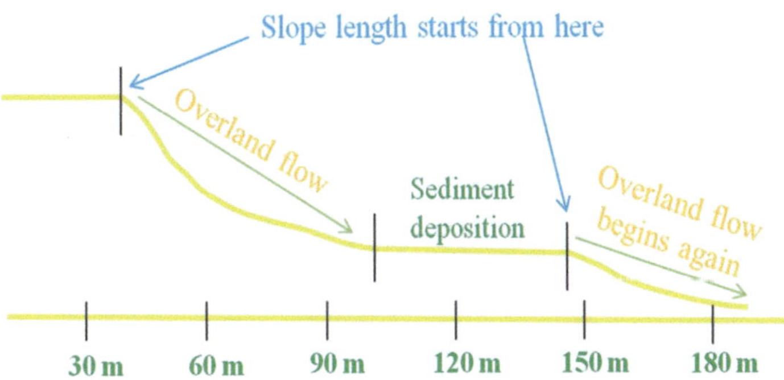

Fig. 4.2 Field estimation of slope length on a side profile of a hill (Khosrowpanah et al. 2007)

4.3.3.2 Slope Gradient Factor *(S)*

Slope gradient factor (*S*) is the ratio of the soil loss from a given gradient of slope to that from land having 9% slope. The slope gradient factor can be expressed mathematically as (Wischmeier and Smith 1978)

$$S = (65.4 \sin^2 \theta + 4.56 \sin \theta + 0.065),\qquad(4.8)$$

where
θ is the angle of slope in degree, which is illustrated in Fig. 4.3.

Then, from Fig. 4.3,

$$\text{Slope (degree)} = \tan^{-1}(\text{rise}/\text{run}).$$

Multiplying the above two factors, the topographic factor (*LS*) is calculated. Thus the topographic factor (*LS*) is the expected ratio of soil loss per unit area from a field slope to that from a 22.13 m length of uniform 9% slope under otherwise identical conditions.

In the present study, the *LS* factor was calculated by an executable C++ program (Khosrowpanah et al. 2007). The program is provided by the International Association of Mathematical Geosciences (IAMG). The C++ program was downloaded from the website http://www.iamg.org.

Khosrowpanah et al. (2007) describes post-processing of the downloaded C++ program. After downloading and uncompressing the package, the C++ executable program along with the source code files is accessible. The DEM input needs to be in ASCII text format to run the program. ArcMap has the function to do this located under the conversion toolbox extension. After the conversion, double-click on the C++

Fig. 4.3 Slope (degrees) shown with the fundamental relationship of a right triangle

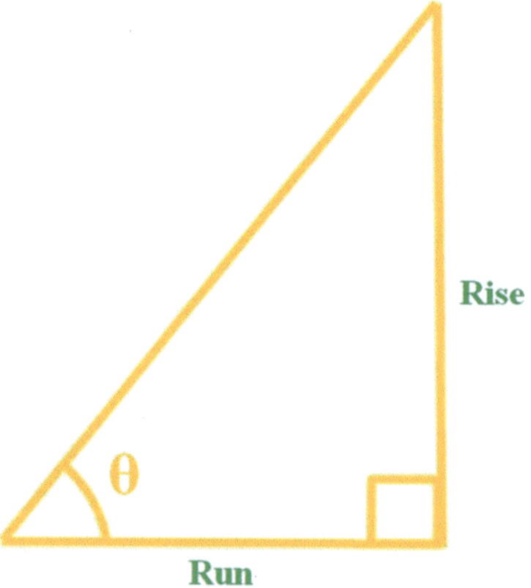

executable program to run it. A series of command lines appear. The first line asks the user to enter the path and filename of the DEM data, which must also be in ASCII text format. Enter the full path to the text file in the form of "E:\Folder name\name of the file with .txt suffix". The second line then asks for the path where all the output files should be placed. Specify this path to lead to the appropriate folder. The third line asks the user to enter a short prefix for the output files. The prefix should not be longer than four letters. The fourth line asks if intermediate files should be produced during the computation process. Select "YES" to see each intermediate output file. The final line then asks if cells with no data should be fixed. The user should select "YES". The program then begins its computation of the DEM text file. The output of program consists of 16 total files with the .dat file suffix. The file suffix must be .txt for the ArcMap to recognize it to convert the output files back to a raster format. Open ArcMap and select the Conversion toolbox. Individually convert the output files to import the LS factor as a raster layer.

4.3.3.3   Description of C++ Program's Operation

A brief description of the C++ executable program's operation is given here. The program begins with a fill function on any depressions or sinks found on the DEM input. The program then identifies the highest elevation points of the DEM, and the flow direction is determined. Conceptually, if rainfall lands on a high point, the direction of flow can be in either one of the cardinal directions (i.e., N, S, E, and W) or the diagonal directions (i.e., NE, SE, SW, and NW). In situations of converging flow, the flow direction of steepest descent takes precedence. The C++ program then calculates the distance between the centers of one grid cell to the next grid cell as the Non-cumulative Slope Length (NCSL). The program's logic for calculating the L factor is summarized below (Hickey 2000).

If the cell being calculated is a high point

$$\text{Then NCSL} = 0.5 \, (\text{cell resolution size})$$

If the input cell's flow direction is in a cardinal (N, S, E, and W) direction

$$\text{Then NCSL} = (\text{cell resolution size})$$

Otherwise (if flow is in diagonal direction: NE, NW, SE, and SW)

$$\text{NCSL} = 1.4142 \, (\text{cell resolution size})$$

A cumulative slope length is then computed by summing the NCSL from each grid cell, beginning at a high point and moving down along the direction of steepest descent. One important part of the C++ program is to recognize the areas where the deposition is the dominant process instead of erosion. The assumption is that deposition would begin in areas where the slope angle decreases sufficiently such

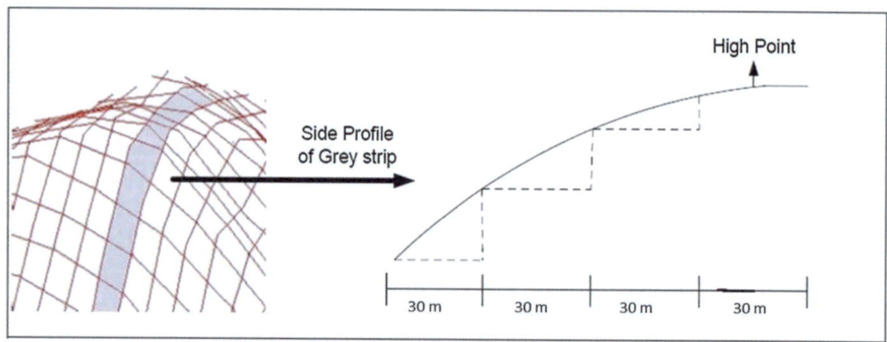

Fig. 4.4 Illustration of the *LS* factor algorithm (Van Remortel et al. 2004)

that overland flow can no longer transport sediment. The program has a function called the cutoff slope angle which is defined as the ratio of change in slope angle from one grid cell to the next along the flow direction. The default values for the slope cutoff angle are 0.5 for slope gradients greater than 5% and 0.7 for slope gradients of less than 5%. These values are based on observations that deposition is easier to initiate on slopes with low gradients (Van Remortel et al. 2004). When the slope angle decreases sufficiently, the cumulative slope length calculation stops. If the land surface extends further downhill, the calculations begin again. The C++ executable program applies the *LS* factor equations (4.7 and 4.8) to each grid cell of the DEM input. Fig. 4.4 illustrates a square mesh of a hill slope with the flow path length colored in gray with a sketch of the side profile.

The primary files used in the computation of the *LS* factor have the following file suffixes: slp_ang for the slope angle of each grid cell, slp_len for the cumulative slope lengths, and slp_exp, which contain the slope dependent exponent *m* that each grid cell is raised to in the *L* factor equation. Recall Eq. (4.7) for calculating the *L* factor:

$$L = \left(\frac{\lambda}{22.13}\right)^{m},$$

where

λ is the horizontally projected slope length (meter), and *m* is the slope length exponent.

The C++ executable program computes the cumulative slope lengths as illustrated in Fig. 4.4 and substitutes this value as λ. After dividing by the reference slope length (22.13 m or 72.6 ft), the expression is raised to the exponent *m*. The slope length exponent *m* is related to the variable β, which is a ratio of rill erosion (defined as erosion caused by overland flow) to inter-rill erosion (defined as erosion caused by rainfall). Equations (4.9) and (4.10) express the method to calculate the exponent *m* (McCool et al. 1987). This equation determines the exponent *m* for areas with an average rill/inter-rill ratio. The low and high rill/inter-rill ratios are obtained by either halving or doubling the value of β, respectively, and substituting them into Eq. (4.9). The C++ executable program uses a low rill/inter-rill ratio in its computation for the exponent m (Van Remortel et al. 2004).

$$m = \frac{\beta}{(\beta + 1)},\tag{4.9}$$

where

m exponent and value of β can be estimated as

$$\beta = \frac{\left(\frac{\sin \theta}{0.0896}\right)}{\left[3.0 \times (\sin \theta)^{0.8} + 0.56\right]},\tag{4.10}$$

where

θ slope angle (degree).

The C++ program first examines the calculated slope angle of each cell and subroutine calls for a table lookup function. The range in which the slope angle falls within is identified, and the corresponding m exponent value is assigned to that cell. For the present study, a geo-coded DEM generated from the Advanced Spaceborne Thermal Emission and Reflection Radiometer (ASTER) data was used. The DEM was downloaded from the GLCF (2000) website with 30-m ground resolution. The DEM of the study area was clipped from downloaded ASTER data. After converting the raster DEM to a program compatible format, the C++ program was run to obtain output files. Later, all the output files were individually converted to raster format using the conversion toolbox of ArcGIS 9.3 to import the LS factor as a raster layer.

4.3.4   Crop/Cover Management Factor (C)

The C factor is the ratio of soil loss from an area with specified cover and management to that from an equal area kept continuously fallow. It measures the effect of canopy and ground cover on the hydraulics of raindrop impact and runoff, as well as the effect of cover and management on the amount and rate of runoff. The C factor represents the effect of plants, soil cover, belowground biomass, and soil-disturbing activities on soil erosion (Jones et al. 1996).

According to current demands for sustainable production, erosion by water should be considered a priority, since it is the consequence of inappropriate land use (Desilva et al. 2010). At the same time, the vegetative soil cover highly contributes to the protection of top soils against erosive processes (Saumer et al. 2010). At the landscape scale, quantities of soil erosion can be modeled with the well-established and worldwide-applied empirical USLE (Wischmeier and Smith 1965; Renard et al. 1997). The vegetative soil cover is considered in the USLE and expressed as the crop/cover management factor (C). Its values range between 0 (very high crop cover protecting the topsoil against soil erosion) and 1 (no effect of the crop cover and high soil loss comparable to that of bare soil) (Saumer et al. 2010).

However, the assessment of this factor to calculate the given soil loss rates is very time-consuming, especially in mountainous areas where access to terrain is often limited. Thus, substantial efforts have been made to calculate and map the C factors for use in soil erosion modeling by means of GIS, remote sensing data, and spectral indices (Asis and Omasa 2007; Suriyaprasit and Shrestha 2008).

Remote sensing techniques are employed to monitor and map the state of ecosystems. Vegetation cover is one of the most important biophysical indicators of soil erosion. Vegetation cover can be estimated using vegetation indices derived from satellite images. Vegetation indices allow us to delineate the distribution of vegetation and soil based on the characteristic reflectance patterns of green vegetation. The Normalized Difference Vegetation Index (NDVI), one of the vegetation indices, measures the amount of green vegetation. Nowadays, the NDVI is used to effectively estimate the crop management factor (C) (Karaburun 2010). The spectral reflectance difference between Near Infrared (NIR) and red is used to calculate NDVI. The equation can be expressed as (Jensen 2000)

$$\text{NDVI} = (\text{NIR} - \text{Red})/(\text{NIR} + \text{Red}). \qquad (4.11)$$

The NDVI has been widely used in remote sensing studies since its development (Jensen 2005). NDVI values range from -1.0 to 1.0, where higher values are for green vegetation and low values are for other common surface materials. Bare soil is represented with NDVI values which are closest to 0, and water bodies are represented with negative NDVI values (Lillesand et al. 2004). More than 20 vegetation indices have been proposed and used at present. Since NDVI provides useful information for detecting and interpreting vegetative land cover, it has been widely used in remote sensing studies (Gao 1996; Myneni and Asrar 1994; Sesnie et al. 2008).

The value of the C factor depends on vegetation type, stage of growth, and cover percentage (Gitas et al. 2009). NDVI values correlate with the C factor (De Jong 1994; Tweddales et al. 2000; De Jong and Riezebos 1997). Many researchers used regression analysis to estimate the C factor values for land cover classes in soil erosion assessment (Lin et al. 2002; Symeonakis and Drake 2004; Van der Knijff et al. 2002).

Various studies assume that a linear correlation exists between the NDVI and C factor. Since the C factor values range from 0 for the well-protected soil to 1 for bare soil (Pierce et al. 1986; Vicenta et al. 2007), the C factor values for bare soil and forest land cover are set to 1 and 0, respectively, in the regression analysis. Figure 4.5 shows the graphical representation of the regression equation. The regression line that describes the relationship between C and NDVI values and R shows the correlation coefficient of regression analysis.

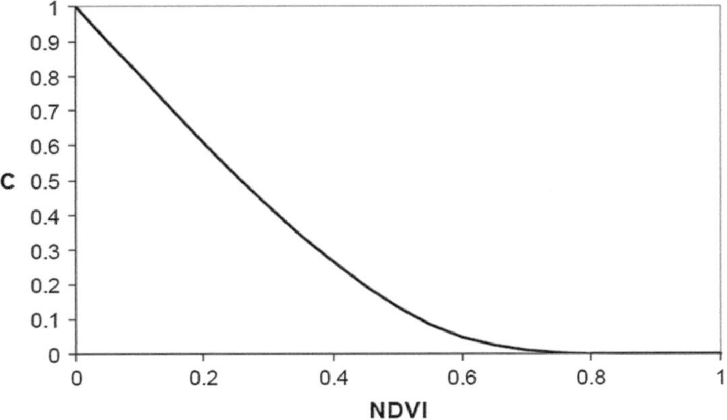

Fig. 4.5   Relation between NDVI and USLE *C* factor

The regression equation is

$$C \text{ factor} = 1.02 - 1.21 \times \text{NDVI}. \tag{4.12}$$

In this study, the land use/land cover map of the study area was prepared. It serves as a guiding tool in the allocation of *C* and *P* factors for different land use/land cover classes (Dabral et al. 2008).

For the preparation of the land use/land cover map, a satellite image of IRS P6 LISS III acquired on January 8, 2011 was used. The image was collected from NRSA, Hyderabad (India). Well-known image processing software ERDAS IMAGINE 9.2 was used to prepare the land use/land cover map. Supervised classification (after several ground truth verifications) with a Maximum Likelihood Classifier (MLC) algorithm was used to classify the satellite image. The error matrix was developed as part of the accuracy assessment to indicate how well the classifier categorized a representative subset of pixels used in the training process of a supervised classification. The estimation based on 938 random samples representing various land use/land cover classes shows an overall accuracy of 89.91%. As per the land use/land cover map, the study area was classified into seven land use/land cover classes, namely agriculture and other vegetation, fallow land, wasteland, forest, habitation, river, and water body. The values of *C* factor suggested by Pandey et al. (2007) were attributed to the land use/land cover map, and a raster map of crop/cover management factor (*C*) was prepared using the conversion toolbox of ArcGIS 9.3.

4.3.5   *Conservation/Support Practice Factor* (P)

The conservation or support practice factor (*P*) is the ratio of soil loss from a land having specified conservation practices to that from a land plowed in a direction parallel to a slope if all other conditions remain unchanged (Wischmeier and Smith 1978).

The support practice factor indicates the rate of soil loss according to the various cultivated lands on the earth. These include contour, cropping, and terrace as its methods, and it is a major factor that can control the erosion (Shin 1999). P values range from 0 to 1, where the value 0 represents an excellent man-made erosion resistance facility and the value 1 indicates no man-made resistance erosion facility (Sheikh et al. 2011).

In this study, the land use/land cover map was also used for the preparation of the P factor map. The values of conservation/support practice factor P as suggested by Dabral et al. (2008) were attributed to the land use/land cover map, and a raster map of the P factor was prepared using the conversion toolbox of ArcGIS 9.3.

4.4 Assessment of Annual Rate of Soil Erosion

The USLE (Wischmeier and Smith 1965) was combined with ArcGIS 9.3 to calculate the average annual rate of soil loss (A) occurring within the Shakkar River watershed. Raster layers corresponding to each of the six USLE factors were created, stored, and analyzed within ArcGIS 9.3. This combination computes the simulated soil erosion potential of the entire watershed, and areas of high soil erosion potential were identified. The grid cells in each layer overlap and the USLE computations were carried out by multiplying all the USLE factors together. Using USLE, the annual rates of soil erosion were estimated for 15 years (1992–2006). Subsequently, the annual rates of soil loss were estimated for each grid cell of the watershed, so that the spatial distribution of the annual soil loss could be presented on cell basis (Fig. 4.6).

4.5 Comparison of Simulated and Observed Sediment Loss

The sediment loss estimated by the USLE was compared with the observed sediment loss data of stream gauging stations located at the outlet of the watershed. The USLE model was validated with the data collected for 15 years (1992–2006), and percent deviation of the simulated sediment loss from the observed values was estimated. The under-prediction or over-prediction limits for the USLE model simulation are within 20% of the measured values and are considered an acceptable level of accuracy for the simulations (Bingner et al. 1989). The USLE model was also validated by plotting the simulated values against the observed values, and a line of best fit and high coefficient of determination was obtained.

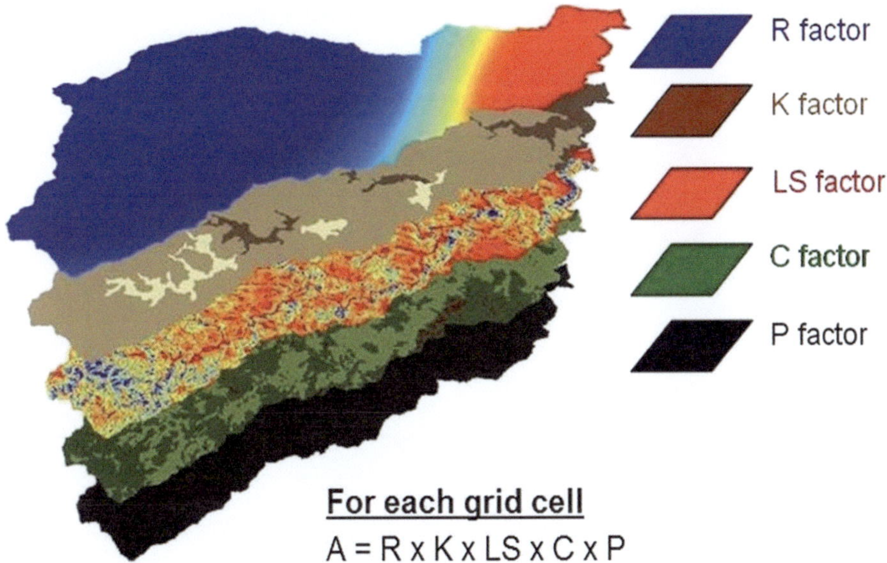

Fig. 4.6 Illustration of the USLE layers and how they overlap (Khosrowpanah et al. 2007)

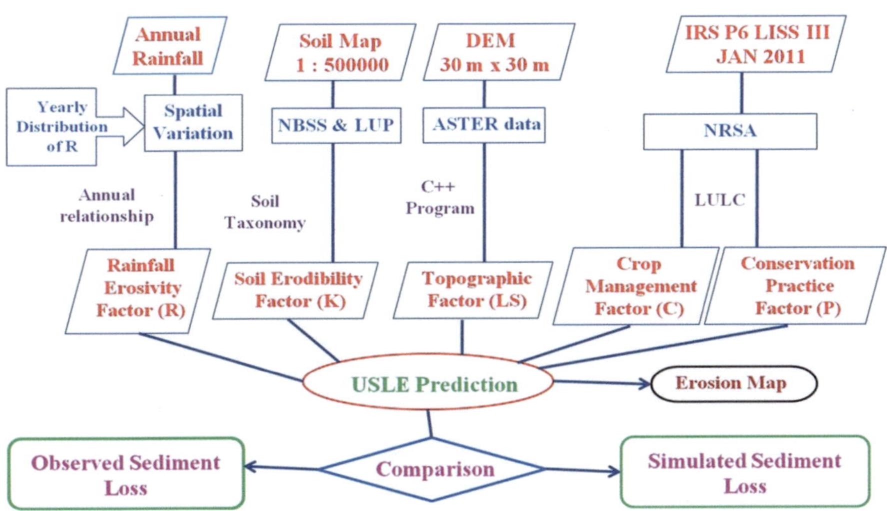

Fig. 4.7 Flowchart for the estimation of the average annual rate of soil loss using USLE and GIS.
© IAHS Press

The detailed methodology for the estimation of the average annual rate of soil loss using the USLE model and its validation with observed data is summarized and presented in Fig. 4.7.

Chapter 5
Distributed Soil Erosion Modeling

Abstract The universal soil loss equation (USLE) and ArcGIS 9.3 were combined to calculate the average annual rate of soil loss (A) within the study area. This combination simulated the soil erosion potential of the entire watershed, and areas of high soil erosion potential were identified. The estimation was carried out for 15 years (1992–2006). The sediment loss estimated by the USLE–GIS combination was compared with the observed sediment loss data from stream gauging station located at the outlet of the watershed. The USLE model was validated with the observed data, and percent deviation of the simulated sediment loss from the observed values was estimated. The annual rates of soil erosion were estimated for a 15-year period (1992–2006) and were found to vary between 6.45 and 13.74 t/ha/year, with an average annual rate of 9.84 t/ha/year. The percentage deviation between simulated and observed values varies between 2.68 and 18.73%, with a coefficient of determination (R^2) of 0.874.

Keywords Rainfall erosivity · Soil erodibility · Topographic factor
Land use land cover · Spatial distribution

Using remote sensing and GIS, spatial erosion modeling by means of the universal soil loss equation (USLE) model was performed to predict the annual rates of soil loss. The model-estimated soil loss was validated with the observed soil loss at the watershed outlet. The results obtained in this study are presented below.

5.1 Preparation of Base Map of the Area

For the delineation of the Shakkar River watershed and the preparation of the drainage map, a geo-coded Digital Elevation Model (DEM) generated from Advanced Spaceborne Thermal Emission and Reflection radiometer (ASTER) data was used. The drainage map of the study area is shown in Fig. 5.1.

The original version of this chapter was revised: Copyright text "© IAHS Press" has been added in the figure and table caption. The erratum to this chapter is available at https://doi.org/10.1007/978-3-319-74286-1_7

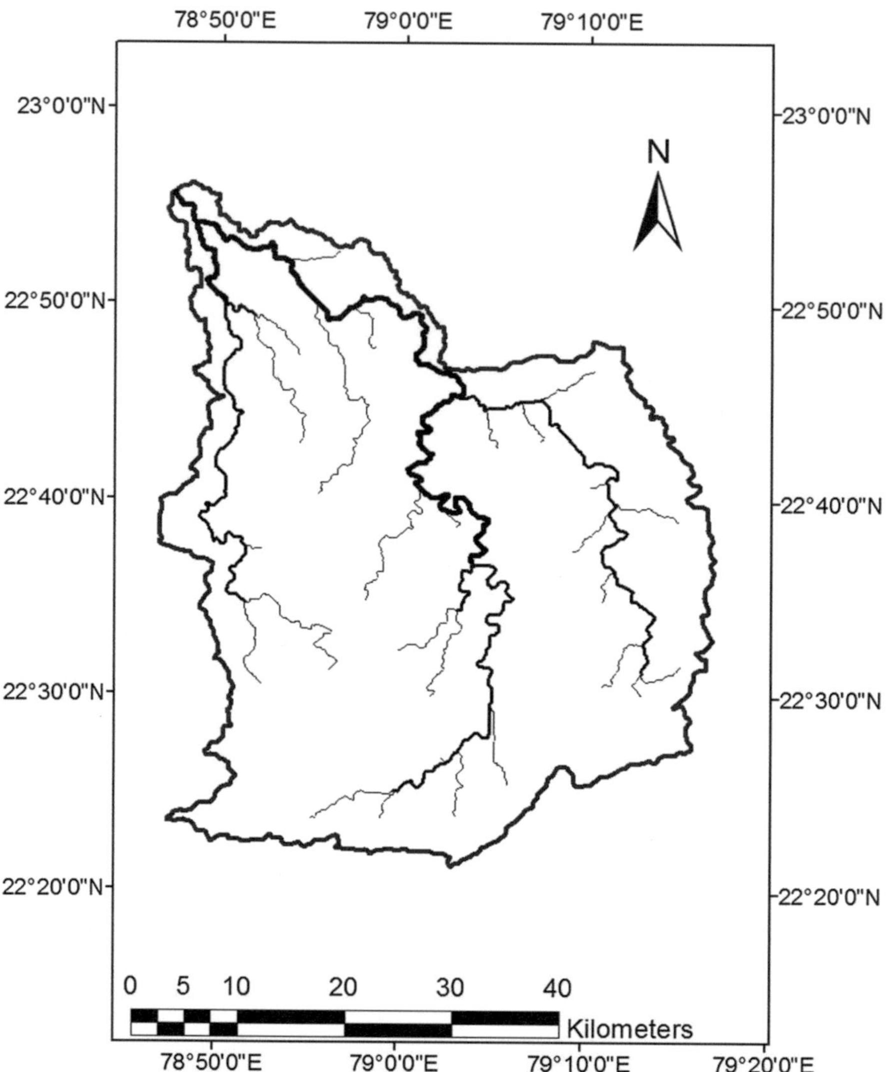

Fig. 5.1 Drainage map of the study area

5.2   USLE-Based Spatially Distributed Soil Erosion Estimation

The rate of soil erosion is strongly dependent upon rainfall, topographic characteristics, soil, and vegetation. These factors vary significantly within a watershed. Therefore, a method, which takes these factors into account while estimating soil erosion, is expected to produce realistic estimates of soil erosion (Das 2008).

The universal soil loss equation USLE (Wischmeier and Smith 1965) is one such equation that takes factors such as rainfall, topography, soil, and land use into consideration while assessing soil erosion.

5.3   Development of Database for USLE

5.3.1   Rainfall Erosivity Factor (R)

Erosion by fluvial processes is believed to be the single most important factor shaping the earth's subaerial landscape. The general factors influencing erosion are well known, but erosive properties of rain play a vital role in soil detachment by impact action (Phillips 1990). The rainfall erosivity factor (R) values for the study area were calculated using regression equation (4.4) following Babu et al. (2004) for Indian catchments. The R factor values calculated for three rain gauge stations are presented in Table 5.1.

For the preparation of the R factor map, a Thiessen polygon of the area was generated by considering three rain gauge stations, as described earlier. The Thiessen polygon of the area is presented in Fig. 5.2.

Then, the above-estimated values of the R factor were attributed to the respective polygon of each rain gauge station. Raster maps of the R factor were generated for

Table 5.1 Annual rainfall and annual rainfall erosivity factor (R)

Station name	Year	Annual rainfall erosivity factor (R)	Average annual rainfall erosivity factor (R)
Gadarwara	1992	426.77	428.54
	1993	567.48	
	1994	408.41	
	1995	359.85	
	1996	293.62	
	1997	442.69	
	1998	231.60	
	1999	921.45	
	2000	434.22	
	2001	458.80	
	2002	473.28	
	2003	474.00	
	2004	419.32	
	2005	567.41	
	2006	341.42	

(continued)

Table 5.1 (continued)

Station name	Year	Annual rainfall erosivity factor (R)	Average annual rainfall erosivity factor (R)
Harrai	1992	467.28	540.16
	1993	566.42	
	1994	763.37	
	1995	442.73	
	1996	398.31	
	1997	627.18	
	1998	465.49	
	1999	850.32	
	2000	468.23	
	2001	517.97	
	2002	448.20	
	2003	597.05	
	2004	451.20	
	2005	538.11	
	2006	499.92	
Amarwara	1992	402.30	515.08
	1993	559.43	
	1994	663.81	
	1995	493.80	
	1996	423.27	
	1997	805.97	
	1998	541.91	
	1999	732.55	
	2000	384.21	
	2001	319.15	
	2002	446.53	
	2003	437.98	
	2004	309.88	
	2005	610.42	
	2006	594.69	

individual years. The R factor map of the year 1992 is shown for representation purposes in Fig. 5.3.

5.3.2 *Soil Erodibility Factor* (**K**)

The soil erodibility factor (K) represents the effect of soil properties and soil profile characteristics on soil loss. In practical terms, the soil erodibility factor is the

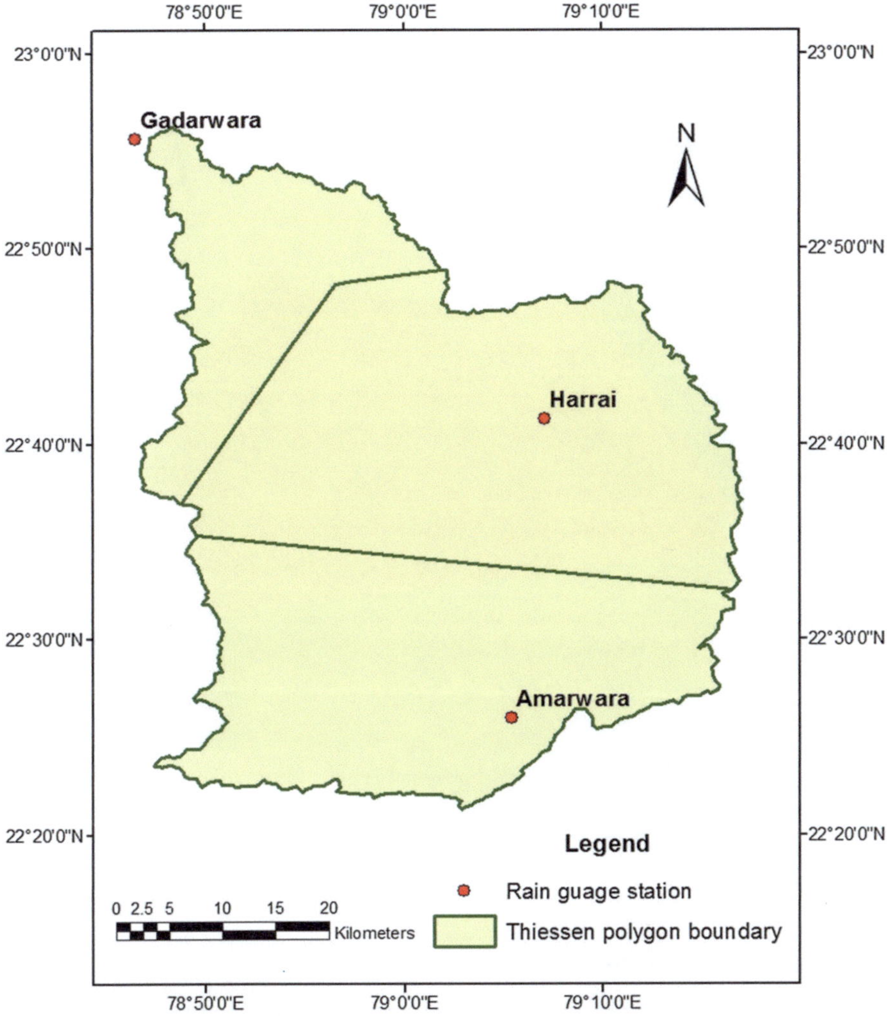

Fig. 5.2 Thiessen polygon of the Shakkar River watershed. © IAHS Press

average long-term soil and soil profile response to the erosive powers of rainstorms. The soil erodibility factor is a lumped parameter that represents an integrated average annual value of the total soil and soil profile reaction to a significant number of erosion and hydrologic processes (Young et al. 1987). The K factor map was generated using the methodology described earlier. The areal extent of the different soil types present in the watershed and assigned values of the K factor for each soil category are given in Table 5.2. Figure 5.4 depicts the spatial distribution of different soil types of the Shakkar River watershed.

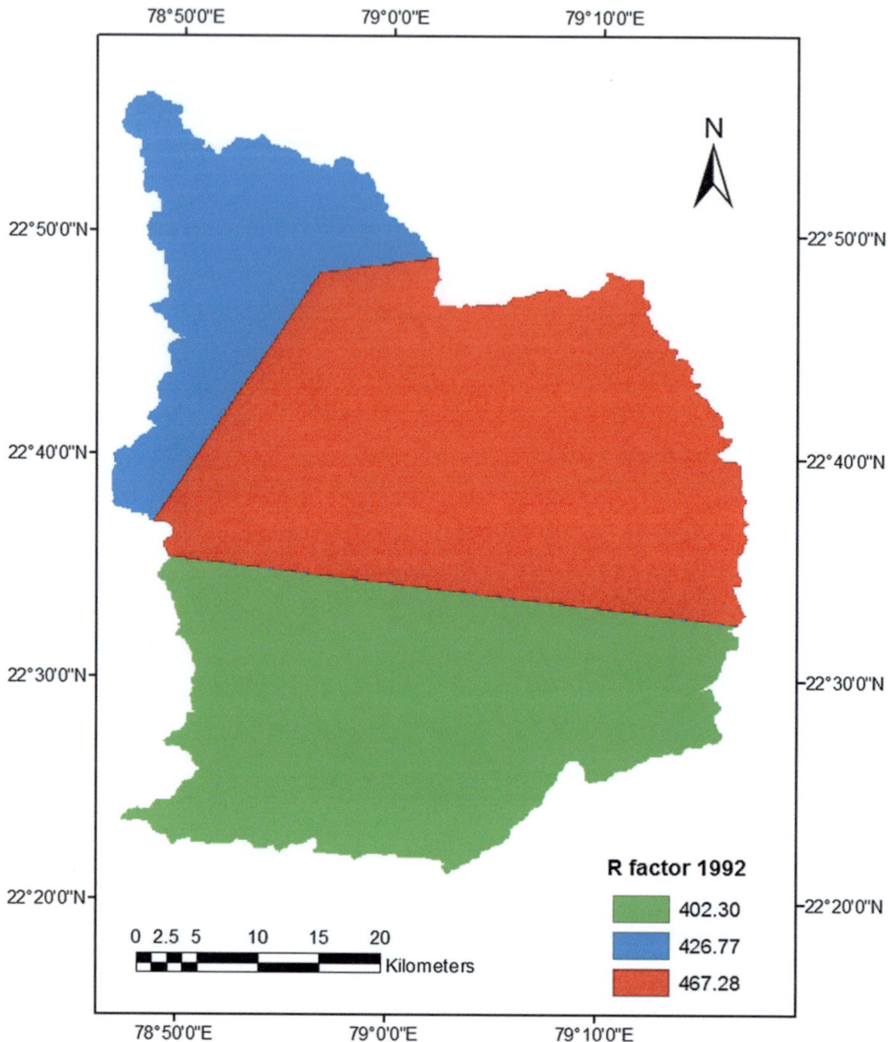

Fig. 5.3 Rainfall erosivity factor (R) map of the year 1992. © IAHS Press

Table 5.2 Area of each soil class, percent of the watershed area, and assigned K factor values. © IAHS Press

Soil type	Soil class area (km^2)	Percent of total watershed area (%)	K factor
Clayey soil	840.12	37.79	0.04
Loamy skeletal soil	95.50	4.29	0.06
Silt loam soil	939.31	42.25	0.07
Loamy soil	318.61	14.33	0.08
Clay loam soil	29.56	1.32	0.11

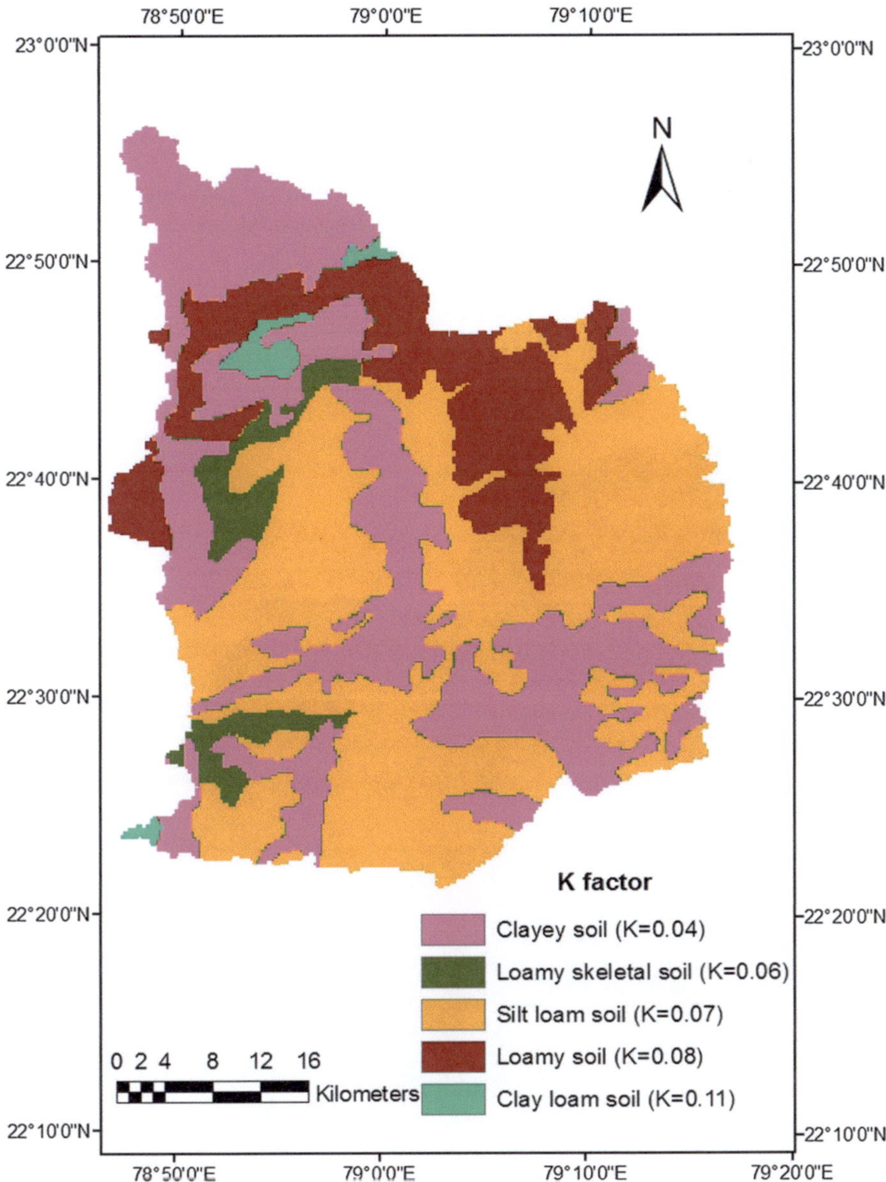

Fig. 5.4 Soil erodibility factor (*K*) map of the Shakkar River watershed. © IAHS Press

5.3.3   *Topographic Factor* (LS)

To compute the *LS* factor, a digital elevation model (DEM) of the area was used. The digital elevation model (DEM) of the study area after filling the sinks is presented in Fig. 5.5.

The topographic factor (*LS*) was calculated by using an executable C++ program. For this, the slope map of the area was prepared and using this map,

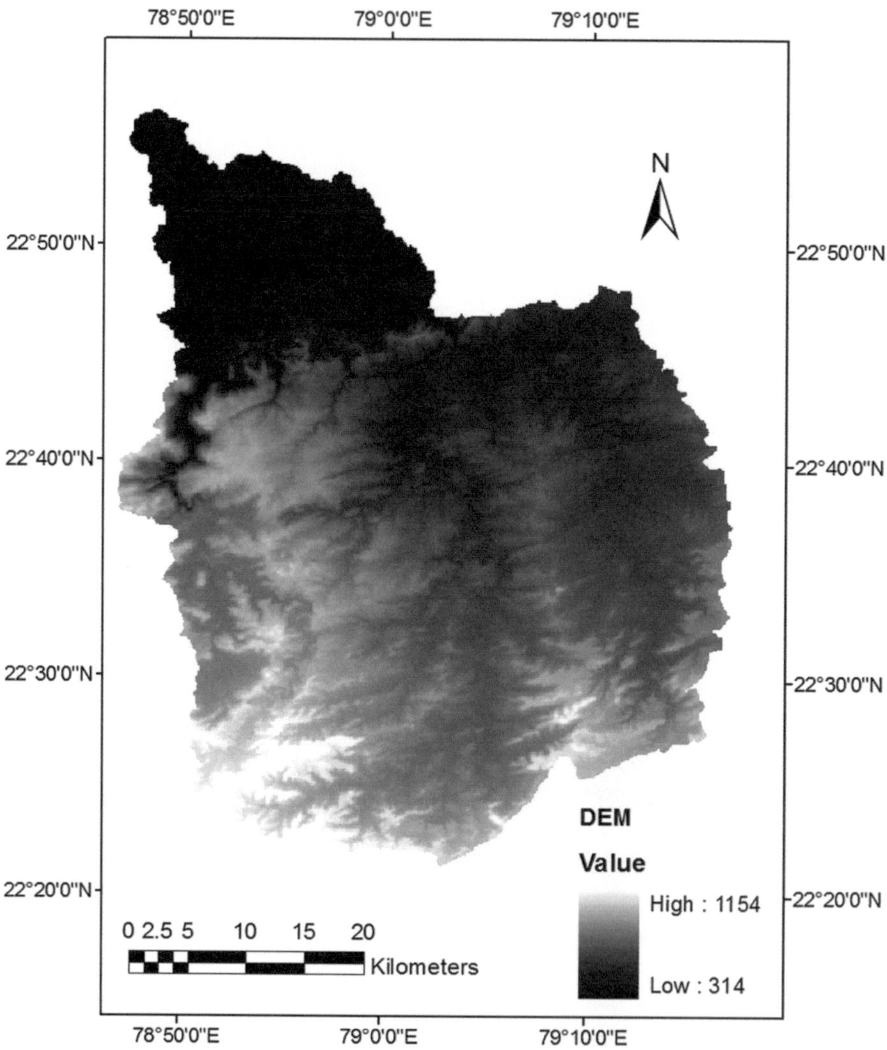

Fig. 5.5 Digital elevation model (DEM) of the study area. © IAHS Press

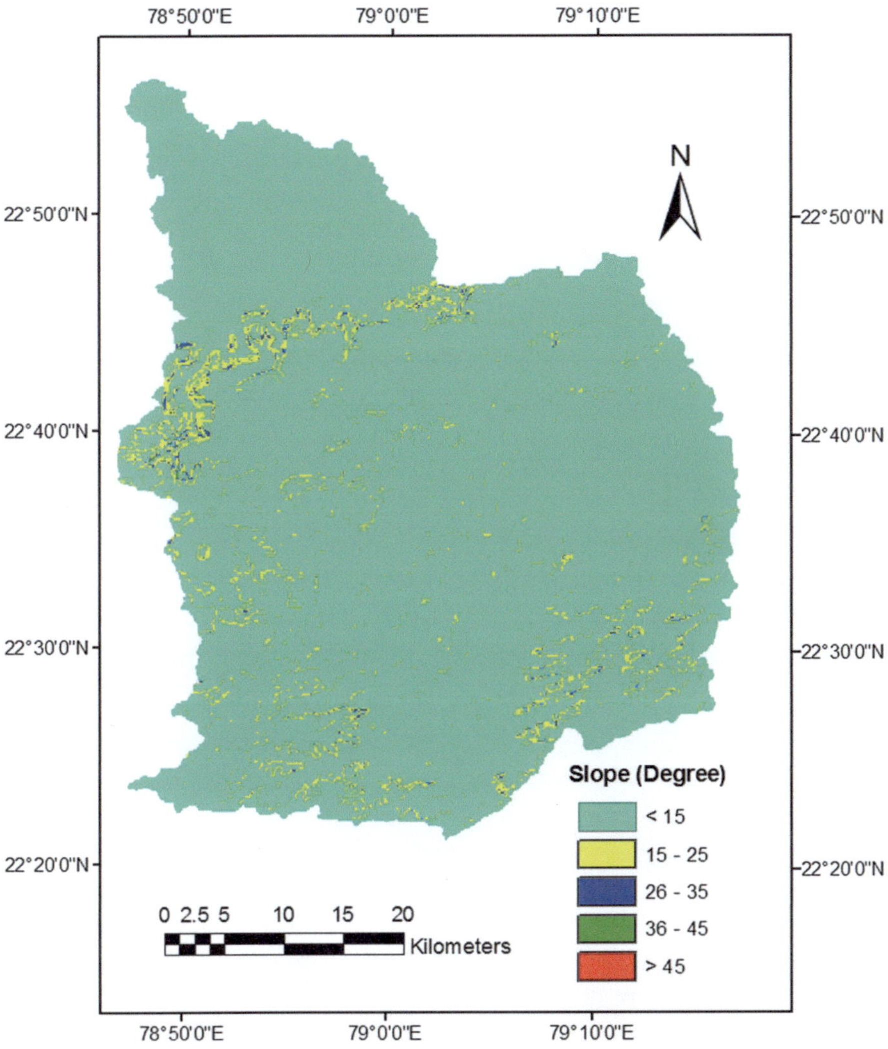

Fig. 5.6 Slope map of the Shakkar River watershed. © IAHS Press

the gradient of the slope factor (S) map and length of slope factor (L) map were prepared. The LS factor map was obtained by multiplying the L factor and S factor maps in GIS environment. The LS factor for the study area was found to vary in the range of 0.1–90.98. The slope map of the watershed is presented in Fig. 5.6 and the LS factor map is presented in Fig. 5.7.

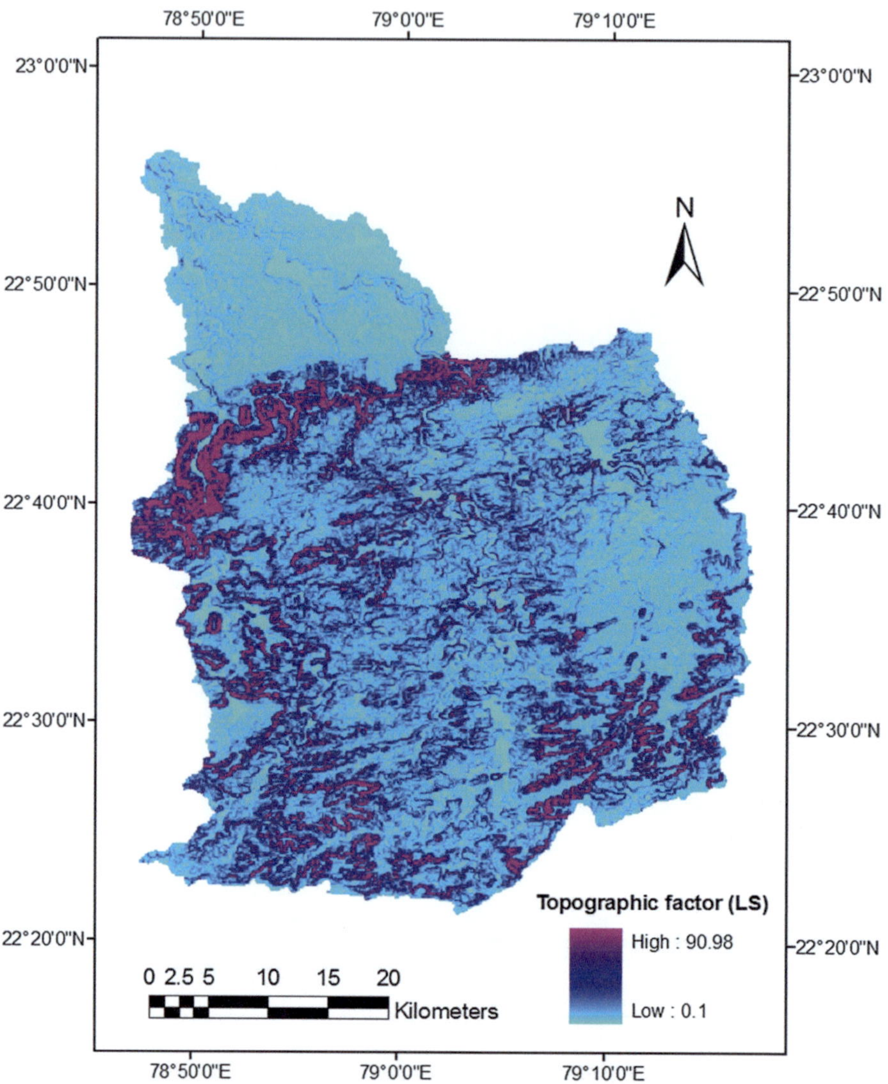

Fig. 5.7 Topographic factor (*LS*) map of the study area. © IAHS Press

5.3.4 Crop/Cover Management Factor (C)

The *C* factor is the ratio of soil loss from a field with specified cover and management to that from an equal area kept continuously fallow. The *C* factor represents the effect of plants, soil cover, below-ground biomass, and soil-disturbing activities on soil erosion (Jones et al. 1996). In the present study, the land use/land cover map was derived from satellite imagery as discussed in the previous chapter. It serves as a guiding tool in the allocation of the C and P factors for different land use classes (Dabral et al. 2008). The C factor values are the representative values for allocating the USLE crop/cover management factors corresponding to each crop/ vegetation condition.

5.3.4.1   Land Use/Land Cover Classification of the Study Area

To obtain information on the vegetation or ground cover, the land use/land cover map of the study area was prepared with the help of ERDAS IMAGINE 9.2 software. As per the land use/land cover classification, the study area was classified into seven land use/land cover classes, namely (1) agriculture and other vegetation, (2) fallow land, (3) wasteland, (4) forest, (5) habitation, (6) river, and (7) water body. The land use/land cover map of the area is presented in Fig. 5.8. The land use/land cover statistics of the study area is presented in Table 5.3.

Finally, values of crop/cover management factor (C) suggested by Pandey et al. (2007) were assigned to different land use land cover patterns and are given in Table 5.4. The magnitude and spatial distribution of the crop/cover management factor is shown in Fig. 5.9. The crop/cover management factor was found to be in the range of 0.004–1.00.

5.3.5   *Conservation/Support Practice Factor* (P)

The conservation or support practice factor (P) is the ratio of soil loss from a land with specified conservation practices to that from a land plowed in a direction parallel to slope if all other conditions remain unchanged (Wischmeier and Smith 1978). In the study area, no major conservation practices are followed. Thus, the values assigned to the P factor were 0.12 for the agricultural area as sugarcane is the major crop in the study area, and 1 for the other areas of the land use/land cover map. These values are suggested by Aggarwal et al. (2000) and Dabral et al. (2008). The magnitude and spatial distribution of the P factor in the study area is shown in Fig. 5.10.

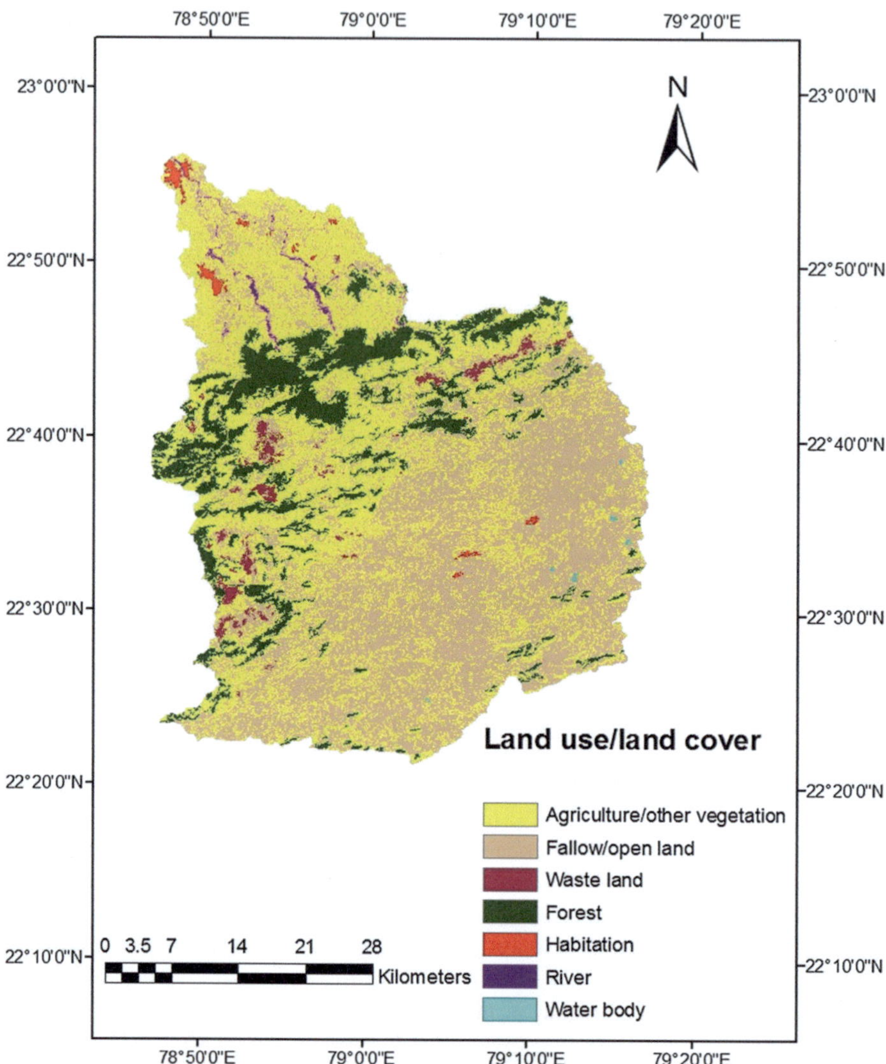

Fig. 5.8 Land use/land cover map of the study area. © IAHS Press

5.4   Assessment of Annual Rate of Soil Erosion

The USLE (Wischmeier and Smith 1965) was integrated with ArcGIS 9.3 to calculate the average annual rate of soil loss (A) occurring within the Shakkar River watershed. Raster layers of rainfall erodibility factor (R), soil erodibility factor (K), topographic factor (LS), crop/cover management factor (C), and conservation/ support practice factor (P) were created, stored, and analyzed within ArcGIS 9.3.

Table 5.3 Land use/land cover statistic of the Shakkar River watershed. © IAHS Press

S. No.	Land use/land cover class	Area (km^2)	Percent of total watershed area (%)
1	Agriculture/other vegetation	723.43	32.54
2	Fallow land	1184.42	53.28
3	Wasteland	32.84	1.47
4	Forest	260.81	11.73
5	Habitation	11.50	0.51
6	River	9.53	0.42
7	Water body	1.10	0.049

Table 5.4 Crop/cover management factor for different land use/land cover classes. © IAHS Press

Land use/land cover class	C factor
Agriculture/other vegetation	0.28
Fallow land	0.18
Wasteland	1
Forest	0.004
Habitation	1
River	0.28
Water body	0.28

This combination computes the simulated soil erosion potential for the entire watershed. Using USLE, the annual rates of soil erosion were estimated for 15 years (1992–2006). The estimated annual soil loss rates are presented in Table 5.5. The annual soil erosion maps generated from remote sensing, GIS, and USLE integration, showing annual rates of soil erosion, are presented in Fig. 5.11 (year 1999, highest rate of soil erosion) and Fig. 5.12 (year 2004, lowest rate of soil erosion).

5.5   Comparison of Simulated and Observed Sediment Loss

The simulated sediment loss and observed sediment loss are presented in Table 5.5.

It is evident from Table 5.5 that the percent deviation of the simulated sediment loss from the observed values varies in the range of 2.68–18.73%. The underprediction or over prediction limits for the USLE model simulation are within 20% of the measured values and are thus considered acceptable levels of accuracy for the simulations (Pandey et al. 2007). The USLE model was also validated by plotting the simulated sediment loss against the observed sediment loss values as shown in

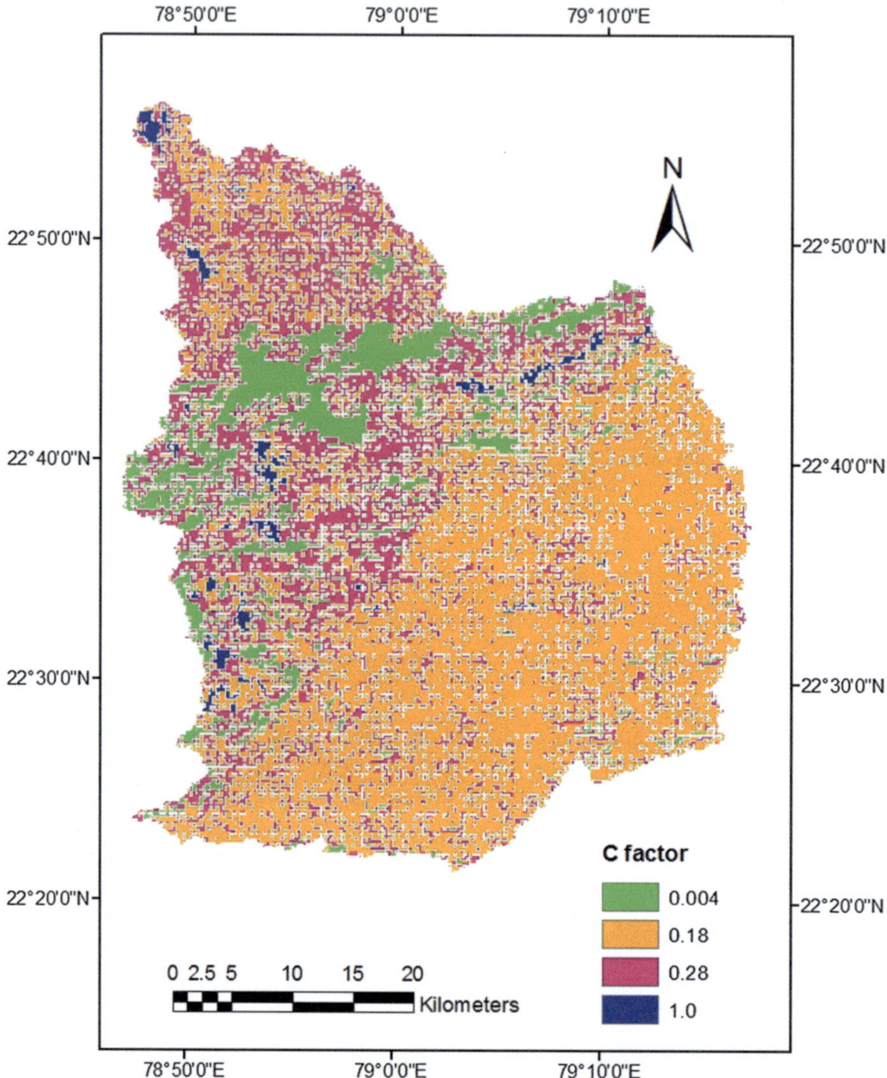

Fig. 5.9 Crop/cover management factor (*C*) map of the study area. © IAHS Press

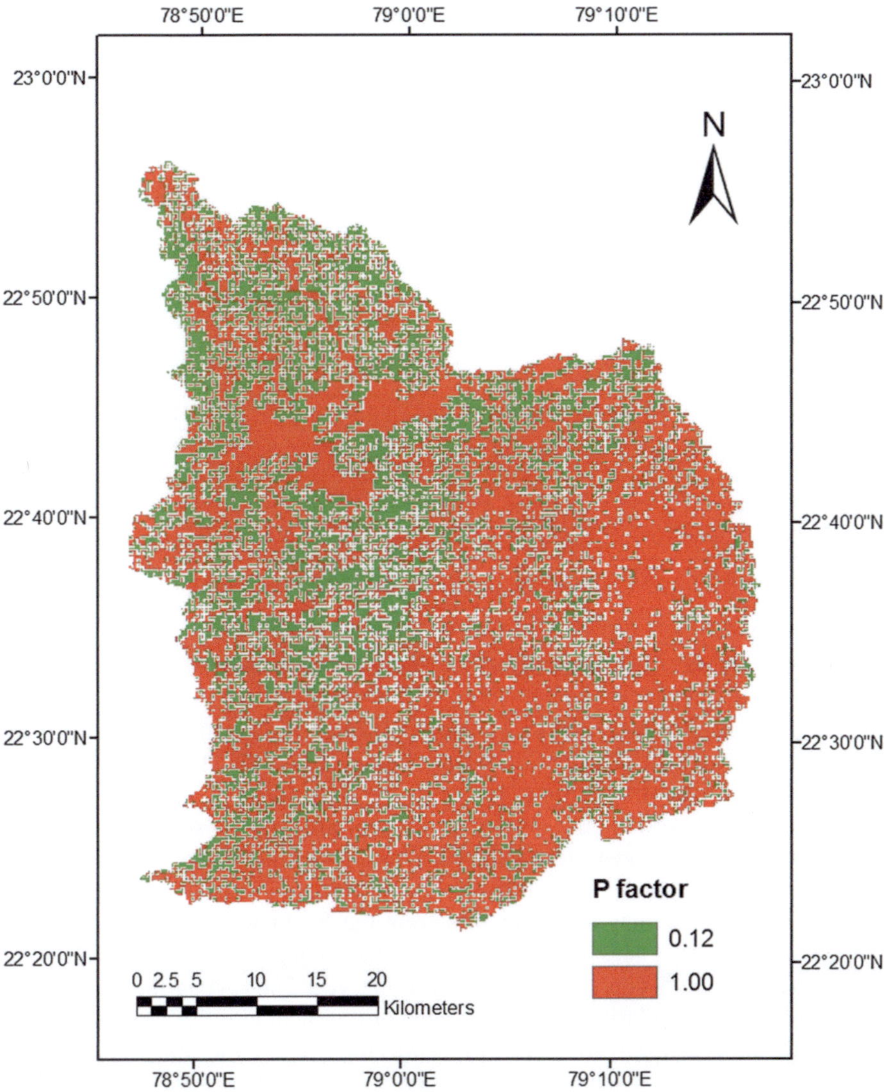

Fig. 5.10 Conservation/support practice factor (*P*) map of the study area. © IAHS Press

Table 5.5 Comparison of observed and simulated sediment losses. © IAHS Press

Year	Observed sediment loss (t/ha/year)	Simulated sediment loss (t/ha/year)	Percent deviation
1992	6.37	7.43	16.64
1993	8.33	9.75	17.05
1994	10.75	11.79	9.67
1995	6.81	8.03	17.91
1996	5.91	6.98	18.10
1997	13.18	12.28	6.83
1998	9.67	8.45	12.62
1999	15.11	13.74	9.07
2000	7.09	7.28	2.68
2001	6.17	7.03	13.94
2002	6.57	7.79	18.57
2003	9.26	8.68	6.26
2004	5.93	6.45	8.77
2005	11.94	10.05	15.83
2006	7.85	9.32	18.73

Fig. 5.13. It can be seen from Fig. 5.13 that the points obtained by plotting the simulated values against observed values are very close to a straight line indicating that the differences between them are not significant. The best-fit line between the above data has a high coefficient of determination (R^2) of 0.874, which shows that they are related by a straight line.

The differences between the observed sediment loss and the simulated sediment loss from the USLE model are a reinforcement of the knowledge that the erosion predictions, in general, contain large factors of error. Thus, USLE can be successfully used to estimate sediment loss from the Shakkar River watershed. The variation in sediment loss is mainly due to the change in R factor. Minimum sediment loss of 6.45 t/ha/year was found in the year 2004 when the rainfall was less than average. Similarly, the highest value of sediment loss, 13.74 t/ha/year was obtained in the year 1999 when the rainfall was high. From this study, the potential soil erosion for the Shakkar River watershed was calculated as 9.84 t/ha/year, on average. It was observed that there is a correlation between the rainfall characteristics and soil loss. An increase in rainfall intensity and amount is accompanied by an increase in soil loss.

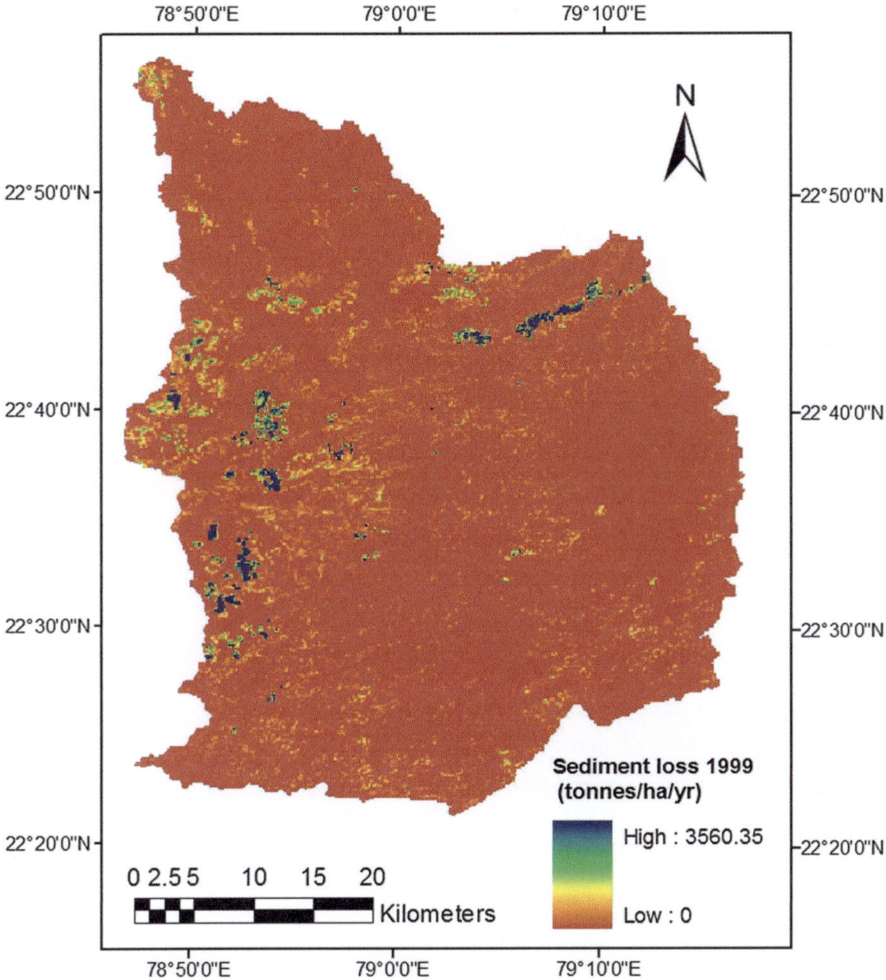

Fig. 5.11 Soil erosion map for the year 1999

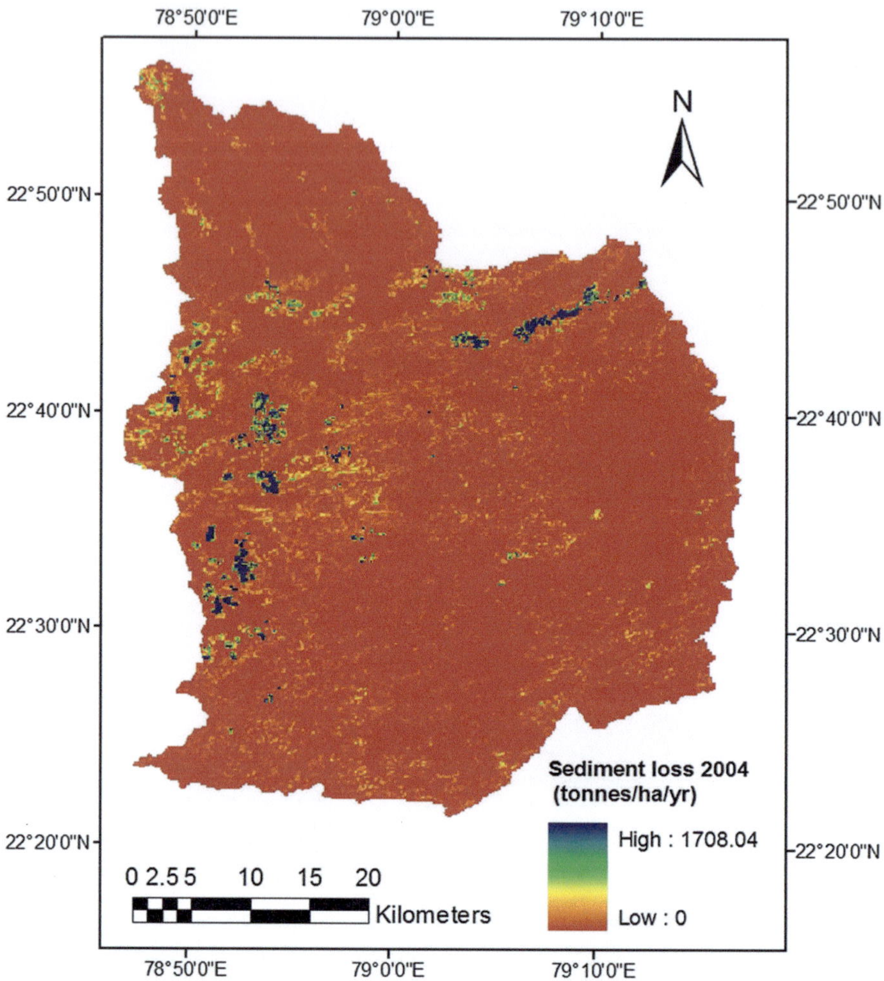

Fig. 5.12 Soil erosion map for the year 2004

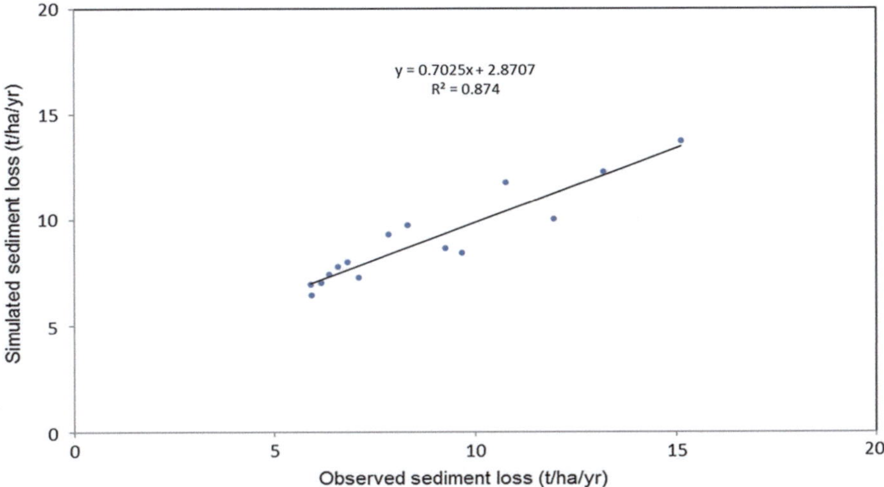

Fig. 5.13 Comparison between observed and simulated sediment loss. © IAHS Press

Table 5.6 Priority scales for soil erosion

Sediment loss (t/ha/year)	Soil erosion class
0–5	Slight
5–10	Moderate
10–20	High
20–40	Very high
40–80	Severe
>80	Very severe

Finally, average annual sediment loss was estimated on a cell basis, and all the grid cells of the watershed were grouped into the following scales of priority as given in Table 5.6.

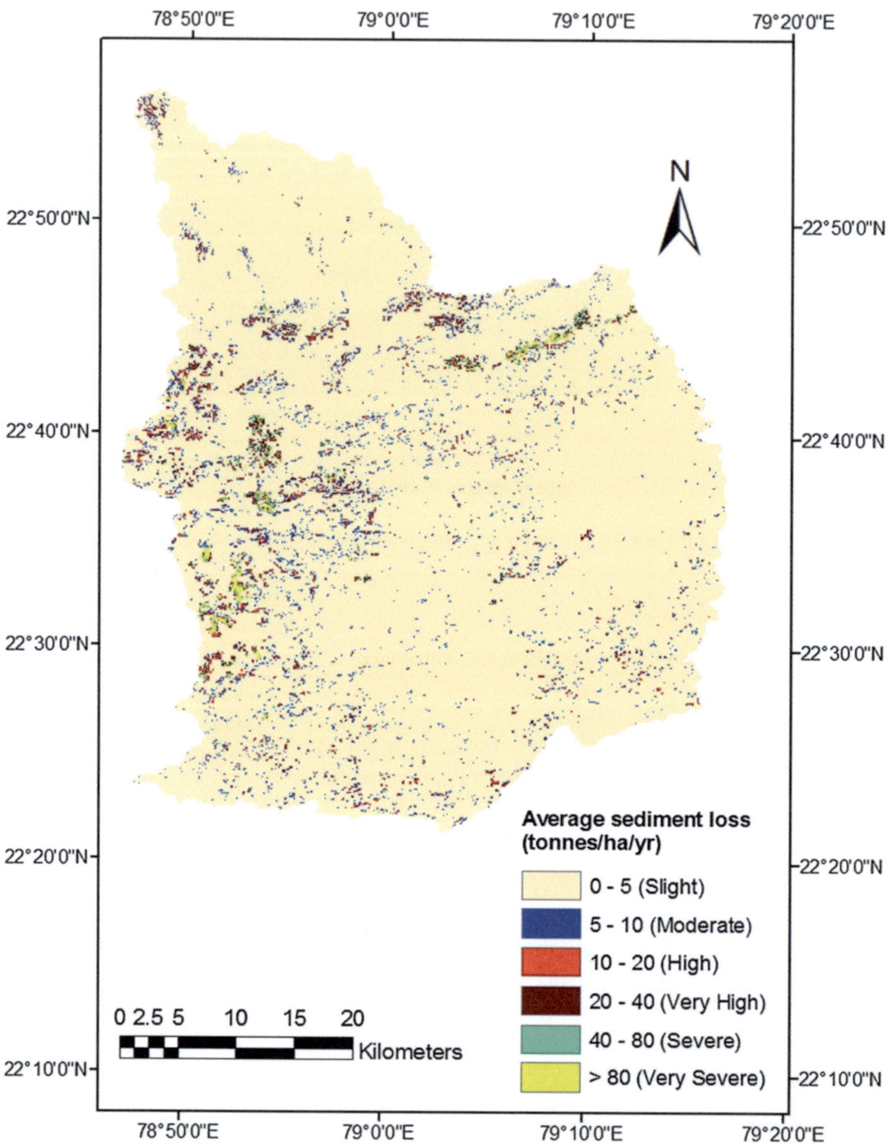

Fig. 5.14 Spatial distribution of average sediment loss in the Shakkar River watershed. © IAHS Press

This classification is based on the guidelines suggested by Singh et al. (1992) for Indian conditions. Fig. 5.14 shows the magnitude and the spatial distribution of potential soil erosion in the Shakkar River watershed on a grid-cell basis.

Chapter 6
Summary, Conclusions, and Suggestions for Further Work

Abstract This chapter summarizes the importance of the spatially distributed soil erosion modeling carried out in this study. Sustainable use of land depends upon the conservation and potential use of soil and water resources. The assessment of the risk of erosion can be helpful for land evaluation. The present study was undertaken to assess the annual rates of soil erosion from a watershed using distributed information for topography, land use, and soil with RS and GIS techniques and to compare the simulated sediment loss with observed sediment loss. The study was conducted in the Shakkar River watershed. The methodology adopted in this study involves the preparation of thematic maps of USLE factors (R, K, LS, C, and P) which are combined in a GIS environment to assess the annual rates of soil erosion. Finally, the simulated sediment loss values were compared with the observed values. This study also demonstrates the scope for future research in distributed soil erosion modeling.

Keywords Soil erosion · Soil and water resources · Distributed soil erosion modeling · Soil erosion risk assessment

6.1 Summary

Soil resources are valuable for the livelihoods of human beings. Sustainable use of land depends on the conservation and potential use of soil and water resources. This severely affects global food security due to an ever-increasing population whose livelihoods depend on limited natural resources. Landslides, mudslides, the collapse of man-made terraces, soil loss from steep slopes, and the decline of forest/pasture areas are the main reasons for land resource degradation. Soil erosion by water is the process of detachment or entrainment, transportation of surface soil particles from its original location and accumulation of it in a new depositional area.

Soil erosion is a global environmental crisis that threatens the natural environment and agriculture. Soil erosion in India is one of the prime concerns of the nation

R. J. Patil, *Spatial Techniques for Soil Erosion Estimation*, SpringerBriefs in GIS, https://doi.org/10.1007/978-3-319-74286-1_6

as agriculture is being adversely affected. Sediment is widely responsible for not only compromising the amount and quality of the water, but also silting the water bodies.

The assessment of the risk of erosion can be helpful for land evaluation in a region where erosion is a major threat to sustained agriculture. Erosion models are used to predict rates of soil erosion. Most of these models need information related to soil type, land use, landform, climate, and topography to estimate soil loss. One of the biggest problems in testing these models is the generation of input data. Conventional methods proved to be too costly and time-consuming for generating this input data. With the advent of remote sensing technology, deriving the spatial information on input parameters has become more feasible and cost-effective. Due to the powerful spatial processing capabilities of geographic information system (GIS) and its compatibility with remote sensing data, soil erosion modeling approaches have become more comprehensive and robust.

The present study was undertaken to assess the annual rates of soil erosion from a watershed using distributed information for topography, land use, and soil with remote sensing (RS) and GIS techniques and to compare the simulated sediment loss with observed sediment loss.

In the present study, the Shakkar River watershed, lying in Narmada River basin situated in Narsinghpur and Chhindwara districts of Madhya Pradesh, was selected. The study area lies between 22° 20′N and 23° 00′N latitudes and from 78° 40′E to 79° 20′E longitudes with an elevation ranging from 314 to 1154 m above mean sea level (MSL). The watershed covers 2223 km^2 of the total geographical area up to the gauging point. The methodology adopted in this study involves the preparation of thematic maps of USLE factors (R, K, LS, C, and P) which are combined in a GIS environment to assess the annual rates of soil erosion. Finally, the simulated sediment loss values were compared with the observed values.

6.2   Conclusions

Based on this study, the following conclusions can be drawn:

1. The annual rates of soil erosion occurring in the study area were estimated by integrating RS and GIS with soil erosion prediction model, USLE. The rates of soil erosion were estimated for 15 years (1992–2006). During this period, the minimum sediment loss of 6.45 t ha^{-1} year^{-1} was found in the year 2004 when the rainfall was lower than average. Similarly, the highest value of sediment loss of 13.74 t ha^{-1} year^{-1} was obtained in the year 1999 when the rainfall was high. The variation in sediment loss is mainly due to the change in R factor. From this study, the potential soil erosion for the Shakkar River watershed was calculated as 9.84 t ha^{-1} year^{-1} on average. It can be concluded that there is a correlation between the rainfall characteristics and soil loss.

2. The simulated sediment loss obtained from USLE was compared with the observed values. The percent deviation of the simulated sediment loss from the observed values varies in the range of 2.68–18.91%. The USLE model was also validated by plotting the simulated sediment loss values against the observed values, and it was found that the points obtained are very close to a line of best fit, indicating that their differences are not significant. The best-fit line between the simulated and observed soil loss has a high coefficient of determination of 0.874, which shows that they are related by a straight line. Thus, USLE can be successfully used for further estimation of sediment loss from the Shakkar River watershed.

Maximum and minimum sediment losses are indications of temporal changes in regional/local rainfall patterns. Looking into the climate change scenarios, sediment loss regimes can be very helpful in establishing links between climate change, land use dynamics, rainfall extremes, runoff generation, and soil erosion. Sediment loss estimates are an important means of comparison between distributed catchments which also help to assess catchment health. As per the soil erosion severity scales in India, most parts of the Shakkar River watershed fall under the slight erosion category and only a very small portion experiences severe erosion. Cropping pattern is the main factor responsible for controlling erosion from cultivable lands, however, illegal activities, mainly stone quarrying in the mountainous parts of the study area, are responsible for the acceleration of soil erosion. In Indian catchments, soil erosion is mainly seasonal in nature as compared to other parts of the world where precipitation is received throughout the year. Southwest monsoon in India mainly dominates the erosion and transport processes. Patterns of southwest monsoon are changing rapidly due to variations in seasonal and annual temperatures. These changes are likely to affect surface processes including soil erosion and may further accelerate the rates of soil erosion.

6.3 Suggestions for Further Work

There is a large scope for future research in this field. Firstly, it is important to carry out distributed soil erosion modeling at different spatial scales/cell sizes to analyze the effect of the spatial scales on soil erosion estimation. Such an analysis is very important in decision-making. This is because the spatial extent of the area does interfere with the cell/grid size used for the distributed modeling as it primarily relates the cost and time required for such projects. Secondly, an elaborative study on the runoff estimation from the watershed should be carried out to validate sediment and runoff estimation models integrated with RS and GIS techniques in the area. Uncertain patterns of rainfall–runoff and gradual increase in the number of extreme rainfall-runoff events as an impact of climate change should be considered when developing rainfall–runoff-based sediment estimation and sediment delivery models.

Erratum to: Spatial Techniques for Soil Erosion Estimation

Erratum to:
R. J. Patil, *Spatial Techniques for Soil Erosion Estimation*,
SpringerBriefs in GIS,
https://doi.org/10.1007/978-3-319-74286-1

In the original version of the book, the following corrections have been incorporated:

Copyright text "© IAHS Press" has been added in the captions of Figures 4.7, 5.2–5.10, 5.13 and 5.14 and in the captions of Tables 5.2–5.5.

"Acknowledgement" has been updated with the inclusion of new text and reference.

The original version has been updated with corrections listed above.

The updated online version of this chapter can be found at
https://doi.org/10.1007/978-3-319-74286-1,
https://doi.org/10.1007/978-3-319-74286-1_4 and
https://doi.org/10.1007/978-3-319-74286-1_5

References

Alejandra MR (2008) Soil erosion calculation using remote sensing and GIS in Río Grande De Arecibo watershed, Puerto Rico. ASPRS, annual conference, Portland, Oregon

Alejandro MA, Omasa K (2007) Estimation of vegetation parameter for modeling soil erosion using linear spectral mixture analysis of Landsat ETM data. ISPRS J Photogram Remote Sens 62:309–324

Amsalu T, Mengaw A (2014) GIS based soil loss estimation using RUSLE model: the case of Jabi Tehinan Woreda, ANRS, Ethiopia. Nat Resour 5:616–626

Arekhi S, Shabani A, Alavipanah SK (2011) Evaluation of integrated KW-GIUH and MUSLE models to predict sediment yield using geographic information system (GIS) (case study: Kengir watershed, Iran). Afr J Agric Res 6(18):4185–4198

Arekhi S, Bolourani AD, Shabani A, Fathizad H, Ahamdyasbchin S (2012) Mapping soil erosion and sediment yield susceptibility using RUSLE, remote sensing and GIS (case study: Cham Gardalan Watershed, Iran). Adv Environ Biol 6(1):109–124

Asis AM, Omasa K (2007) Estimation of vegetation parameter for modeling soil erosion using linear spectral mixture analysis of Landsat ETM data. ISPRS J Photogram Remote Sens 62 (4):309–324

Avanzi JC, Beskow S, Curi N, Mello CR, Norto LD, Viola MR (2009) Soil erosion prediction in the Grande River Basin, Brazil using distributed modeling. Catena 79:49–59

Babu R, Dhyani BL, Kumar N (2004) Assessment of erodibility status and a refined Iso-erodent map of India. Indian J Soil Conserv 32(3):171–177

Bahrami HA, Vaezi AR, Sadeghi SHR, Mahdian MH (2011) Developing a nomograph for estimating erodibility factor of calcareous soils in North West of Iran. Int J Geol 4(5)

Bahrawi JA, Elhag M, Aldhebiani AY, Galal HK, Hegazy AK, Alghailani E (2016) Soil erosion estimation using remote sensing techniques in Wadi Yalamlam Basin, Saudi Arabia. Advances in Materials Science and Engineering. http://dx.doi.org/10.1155/2016/9585962

Batistella M, Lu D, Li G, Valladares GS (2004) Mapping soil erosion risk in Rondo Nia, Brazilian Amazonia: using RUSLE, remote sensing and GIS. Land Degrad Develop 15:499–512

Beasley DB, Huggins LF, Monke EJ (1980) ANSWERS: a model for watershed planning. Trans ASAE 23:938–944

Bhaware KO (2006) Soil erosion risk modeling and current erosion damage assessment using remote sensing and GIS techniques. Master's Thesis, Andhra University, India

Bhunia GS, Pal B, Samanta S (2012) Quantitative analysis of relief characteristics using space technology. Int J Physic Soc Sci 2(8)

Bingner RL, Murphee CE, Mutchler CK (1989) Comparison of sediment yield models on various watersheds in Mississipi. Trans ASAE 32(2):529–534

Cohen MJ, Shepherd KD, Walsh MG (2005) Empirical formulation of the USLE for erosion risk assessment in a tropical watershed. Gesderma 124:235–252

Dabral PP, Baithuri N, Pandey A (2008) Soil erosion assessment in a hilly catchment of North Eastern India using USLE, GIS, and remote sensing. Water Resour Manag 22:1783–1798

Das DJ (2008) Identification of critical erosion prone areas for watershed prioritization using GIS and remote sensing. Master's Thesis, Indian Institute of Technology, Roorkee

Das GS (2010) Hydrology and soil conservation engineering including watershed management, 2nd edn (Eastern Economy Edition). PHI Learning, Private Limited, New Delhi

De Jong SM (1994) Application of reflective remote sensing for land degradation studies in a mediterranean environment. University of Utrecht

De Jong SM, Riezebos HT (1997) SEMMED: a distributed approach to soil erosion modeling. In: Spiteri A (ed) Integrated applications for risk assessment and disaster prevention for the mediterranean. Balkema, Rotterdam, pp 199–204

Dengiz O, Yakupoglu T, Baskan O (2009) Soil erosion assessment using geographical information system (GIS) and remote sensing (RS) study from Ankara-Guvenc Basin, Turkey. J Environ Biol 30(3):339–344

Desilva RP, Dayawansa NDK, Jayarathne KDBL (2010) GIS-based analysis of biophysical and socio-economic factors for land degradation in Kandaketiya DS division. Tropical Agric Res 21(4):361–367

Douros K, Gitas IZ, Karydas CG, Minakou C, Silleos GN (2009) Multi-temporal soil erosion risk assessment in North Chalkidiki using a modified USLE raster model. EARSeL e-Proceedings 8(1)

Dudal R (1981) An evaluation of conservation needs. Soil conservation: problems and prospects. Wiley, New York

Dwivedi RS, Ravi Sankar T, Venkataratnam L, Karale RL, Gawande SP, Seshagin Rao KV, Senchaudhary S, Bhaumik KR, Mukharjee KK (1997) The inventory and monitoring of eroded lands using remote sensing data. Intl J Remote Sens 18(1):107–119

Eiumnoh A (2000) Integration of geographic information system (GIS) and satellite remote sensing (SRS) for watershed management. School of Environment, Resources and Development, Asian Institute of Technology, Thailand

Elangovan AB, Seetharaman R (2011) Estimating rainfall erosivity of the revised universal soil loss equation from daily rainfall depth in Krishanagiri watershed region of Tamil Nadu, India. International conference on environmental and computer science, Singapore, Vol 19

El-Swaify SA (1997) Factors affecting soil erosion hazards and conservation needs for steep tropical lands. Soil Technol 11(1):3–16

ERDAS IMAGINE Tour Guides, ERDAS IMAGINE Field Guides, ERDAS IMAGINE V 8.5 (2001) Inc. Atlanta, Georgia, USA

ESRI (Environmental Systems Research Institute) (1994) Cell-based modeling with the grid. Environmental Systems Research Institute Inc., Redlands, California, USA

Esther MW (2009) Using GIS techniques to determine RUSLE's 'R' and 'LS' factors for Kapingazi River catchment. Published Master's research, Jomo Kenyatta University of Agriculture

FAO/UNEP (1994) Land degradation in South Asia: its severity causes and effects upon the people. FAO and UNEP project, Rome

Fernandez CC, Stockle O, Wu JQ, McCool DK (2002) Predicting erosion and sediment yield using geographic information systems: application to the Lawyers Creek watershed. Research and extension regional water quality conference, Washington

Fernandez C, Wu JQ, McCool DK, Stockle CO (2003) Estimating water erosion and sediment yield with GIS, RUSLE, and SEDD. J Soil Water Conserv 58(3):128–136

Fistikoglu O, Harmancioglu NB (2002) Integration of GIS with USLE in the assessment of soil erosion. Water Resour Manag 16:447–467

Fujaco MAG, Leite MGP, Neves AHCJ (2016) A GIS-based tool for estimating soil loss in agricultural river basins. REM, Int Eng J 69(4). http://dx.doi.org/10.1590/037044672015 690197

Gallant J, Lu H, Moran C, Priestley G, Prosser IP (2001) Prediction of sheet and rill erosion over the Australian Continent, incorporating monthly soil loss distribution. CSIRO, Land, and Water, Canberra, Australia, technical report 13(01)

Gao BC (1996) NDWI—A normalized difference water index for remote sensing of vegetation liquid water from space. Remote Sens Environ 58:257–266

Gelagay HS, Minale AS (2016) Soil loss estimation using GIS and remote sensing techniques: a case of Koga watershed, North-western Ethiopia. Int Soil Water Cons Res 4:126–136

Gitas IZ, Douros K, Minakou C, Silleos GN, Karydas CG (2009) Multi-temporal soil erosion risk assessment in North Chalkidiki using a modified USLE raster model. Earsel e-proceedings 8:40–52

Global Land Cover Facility (GLCF 2000) http://glcfapp.umiacs.umd.edu. Visited on 10 Aug 2012

Gunawan G, Sutjiningsih D, Soeryantono H, Sulistioweni W (2013) Soil erosion estimation based on GIS and remote sensing for supporting integrated water resources conservation management. Int J Tech 2:147–156

Hakim AVM, Chandrakaran S (2005) Hydrology of high and midland watersheds of Kerala. Edi. book; Hydrology and Watershed Management, Himanshu Publications, Udaipur, pp 224–233

Hickey R (2000) Slope angle and slope length solutions for GIS. Cartography 29(1):1–8

Hoshikawa K, Kawabata M, Shrestha MB, Suzuki J (2005) Estimation of potential sediment yield by integrating USLE with GIS—a case study at Tenryu watershed in Central Japan. Shinshu University, Nagano, Japan

Htun KZ, Aye SS, Samarakoon L (2007) Spatial pattern analysis of land degradation using satellite remote sensing data and GIS in Mandalay watershed. Mandalay Technological University, Myanmar, Central Myanmar

Hudson N (1995) Runoff, erosion, and sedimentation: prediction and measurement. In: FAO (ed) Land and water integration and river basin management. FAO Land Water Bull 1(85)

Hui L, Huizhong H, Xiaoling C, Lihua Z (2008) An approach to compute the C factor for universal soil loss equation using EOS-MODIS vegetation index (VI). SPIE Digit Lib 7285

IAMG (International Association of Mathematical Geosciences), http://www.iamg.org. Accessed 15 Sept 2012

Integrated Hydrological Data Book (Non-classified river basins) (2006) Published by Hydrological Data Directorate, Information Systems Organization, Water Planning and Projects Wing, Central Water Commission, New Delhi

Integrated Hydrological Data Book (Non-classified river basins) (2012) Published by Hydrological Data Directorate, Information Systems Organization, Water Planning and Projects Wing, Central Water Commission, New Delhi

Jain MK, Kothyari UC (2000) Estimation of soil erosion and sediment yield using GIS. Hydrol Sci J 45(5):771–786

Jain SK, Goel MK (2002) Assessing the vulnerability to soil erosion of the Ukai Dam catchment using remote sensing and GIS. Hydrol Sci J 47(1):31–40

Jensen JR (2000) Remote sensing of the environment: an earth resource perspective. Prentice Hall, New Jersey

Jensen JR (2005) Introductory digital image processing: a remote sensing perspective. Pearson Prentice Hall, New Jersey

Jones RJA, Montanarella L, Van der Knijff JM (1996) Soil erosion risk assessment in Italy. Space Applications Institute, European Soil Bureau

Julien YP, Tanago MGD (1991) Spatially varied soil erosion under different climates. Hydrol Sci J 36(6):511–524

Karaburun A (2010) Estimation of C factor for soil erosion modeling using NDVI in Buyukcekmece watershed. Ozean J Appl Sci 3(1)

Kavian A, Azmoodeh A, Solaimani K (2014) Deforestation effects on soil properties, runoff and erosion in northern Iran. Arab J Geosci 7:1941–1950

Khosrokhani M, Pradhan B (2014) Spatio-temporal assessment of soil erosion at Kuala Lumpur metropolitan city using remote sensing data and GIS. Geomatics Nat Hazards Risk 5(3): 252–270

Khosrowpanah S, Heitz LF, Wen Y, Park M (2007) Developing a GIS-based soil erosion potential model of the Ugum watershed. Water and Environmental Research Institute, Western Pacific UOG Station, Guam, Technical Report No. 117

Kim HS (2006) Soil erosion modeling using RUSLE and GIS on the Imha Watershed, South Korea. Master's thesis, Colorado State University, Fort Collins, Colorado

Kothyari UC, Jain SK (1997) Sediment yield estimation using GIS. Hydrol Sci J 42(6)

Kothyari UC, Jain MK, Ranga Raju KG (2002) Estimation of temporal variation of sediment yield using GIS. Hydrol Sci J 47(5):693–706

Kumar V (2003) Rainfall characteristics of Shimla district (H.P.). J Indian Water Resour Soc 23 (1):1–10

Kumar V, Rastogi RA (2005) Sub-area routing model for estimation of sediment yield from a mountainous watershed. Edi. book: Hydrology and Watershed Management, Himanshu Publications, Udaipur, pp 98–105

Lal R (1998) Soil erosion impact on agronomic productivity and environment quality: critical reviews. Plant Sci 17:319–464

Lal R (2001) Soil degradation by erosion. Land Degrad Develop 12:519–539

Liengsakul M, Mekpaiboonwatana S, Pramjanee P (1993) Use of GIS and remote sensing for soil mapping and for locating new sites for permanent cropland: a case study in the highlands of northern Thailand. Geoderma 60:293–307

Lillesand TM, Kiefer RW, Chipman JW (2004) Remote sensing and image interpretation, 5th edn. Wiley, New York

Lim KJ, Sagong M, Engel BA, Tang Z, Choi J, Kim KS (2005) GIS-based sediment assessment tool. Catena 64:61–80

Lin CY, Lin WT, Chou WC (2002) Soil erosion prediction and sediment yield estimation: the Taiwan experience. Soil and Tillage Res 68(2):143–152

Mali VK, Pathak K, Tiwari KN, Tripathy SK (2016) Remote sensing and GIS based assessment of soil erosion and soil loss risk around hill top surface mines situated in Saranda Forest, Jharkhand. J Water Clim Change 7(1):68–82

Mahmoodabadi M (2011) Sediment yield estimation using a semi-quantitative model and GIS-remote sensing data. Int Agrophysics 25:241–247

Mati BM, Morgan RPC, Gichuki FN, Quinton JN, Brewer TR, Liniger HP (2000) Assessment of erosion hazard with the USLE and GIS: a case study of the Upper Ewaso Ngiro North basin of Kenya. JAG 2(2):78–86

McCool DK, Brown LC, Foster GR (1987) Revised slope steepness factor for the USLE. Trans ASAE 30:1387–1396

Mishra RN, Bhist SS, Chandra Sekhar M (2003) Watershed management—a perspective, Edi. book, Allied Publishers Limited, New Delhi. pp 151–163

Mishra PK, Singh AK, Kothari M, Kumar V, Srinivasa Reddy GV, Purohit RC (2007) Soil erosion research under simulated conditions for black soils of Southern India—a review. Agric Rev 28 (11):63–68

Morgan RPC (1995) Soil erosion and conservation, 2nd edn. Longman Group, UK Limited

Morgan RPC, Morgan DDV, Finney HJ (1984) A predictive model for the assessment of soil erosion risk. J Agril Eng Res 30(1):245–253

Morgan RPC, Morgan DDV, Finney HJ (1999) A predictive model for the assessment of soil erosion risk. Can J Remote Sens 25(4):367–380

Mosbahi M, Benabdallah M, Boussema RM (2013) Assessment of soil erosion risk using SWAT model. Arab J Geosci 6:4011–4019

Myneni RB, Asrar G (1994) Atmospheric effects, and spectral vegetation indices. Remote Sens Environ 47:390–402

NBSSLUP (1996) Soils of Madhya Pradesh for optimizing land use. Publication, 50(b)

Nearing MA, Lane LJ (1989) A process-based soil erosion model for USDA water erosion prediction project technology. Trans ASCE 32(5):1587–1593

Okoth PF (2003) A hierarchical method for soil erosion assessment and spatial risk modeling (a case study of Kiambu District in Kenya). Doctoral Thesis, Wageningen University

Pal B, Samanta S (2011) Estimation of soil loss using remote sensing and geographic information system techniques (case study of Kaliaghai River basin, Purba and Paschim Medinipur District, West Bengal, India). Indian J Sci Technol 4(10)

Pandey A, Chowdary VM, Mal BC (2007) Identification of critical erosion prone areas in the small agricultural watershed using USLE, GIS, and remote sensing. Water Resour Manag 21(4): 729–746

Patil RJ, Sharma SK, Tignath S (2015) Remote sensing and GIS based soil erosion assessment from an agricultural watershed. Arab J Geosci 8:6967–6984

Phadke VS, Singh R (2006) Assessing soil loss by water erosion in Jamni River Basin, Bundelkhand region, India, adopting universal soil loss equation using GIS. Current Sci 90(10)

Phillips JD (1990) Relative importance of factors influencing fluvial soil loss at a global scale. Am J Sci 290:547–568

Pierce FJ, Larson WE, Dowdy RH (1986) Soil conservation: an assessment of the national resources inventory, vol 2. National Academy Press, Washington (DC)

Pradhan B, Chaudhari A, Adinarayana J, Buchroithner MF (2012) Soil erosion assessment and its correlation with landslide events using remote sensing data and GIS: a case study at Penang Island, Malaysia. Environ Monit Assess 184(2):715–727

Renard KG, Foster GR, Weesies GA, McCool DK, Yoder DC (1997) Predicting soil erosion by water: a guide to conservation planning with the revised universal soil loss equation (RUSLE). Handbook, 703, US Department of Agriculture: Washington (DC), 404

Rompaey AV, Bazzoffi P, Jones RJA, Montanarella L (2005) Modeling sediment yields in Italian catchments. Geomorphology 65:157–169

Roslan ZA, Tew KH (1997) Use of satellite imagery to determine the land use management factors of the USLE. In: Human impact on erosion and sedimentation (proceedings of rabat symposium S6), IAHS, Publication no. 245

Rouse JW, Haas RW, Schell JA (1974) Monitoring the vernal advancement and retrogradation (greenwave effect) of natural vegetation. NASA/GSFCT, Type-III final report, Greenbelt, MD, USA

Saha SK, Kudarat M, Bhan SK (1992) Erosion soil loss prediction using digital satellite data and universal soil loss equation—soil loss mapping in Siwalik Hills in India. In Shunji mural (ed) Application of remote sensing in asia and oceanic-environmental change monitoring: Phb. Asian Association on Remote Sensing, pp 369–372

Said BS, Svoray T (2009) Soil loss, water ponding and sediment deposition variations as a consequence of rainfall intensity and land use: a multi-criteria analysis. Earth Surface Processes and Landforms 102(10)

Saumer P, Schonbrodt S, Behrens T, Seeber C, Scholten T (2010) Assessing the USLE crop and management factor C for soil erosion modeling in a large mountainous watershed in Central China. J Earth Sci 21(6):835–845

Schmidt L, Rasemann S, Schrott J, Dikau R (2002) Geomorphometry in mountain terrain. In: Bishop M, Shroder JJ (eds) Geographic information science (GIScience) and mountain geomorphology. Praxis Scientific Publishing, Springer, Berlin

Sesnie SE, Gessler PE, Finegan B (2008) Integrating Landsat TM and SRTM-DEM derived variables with decision trees for habitat classification and change detection in complex neotropical environments. Remote Sens Environ 112:2145–2159

Sheikh AH, Palria S, Alam A (2011) Integration of GIS and universal soil loss equation (USLE) for soil loss estimation in a Himalayan watershed. Recent Res Sci Technol 3(3):51–57

Shin GJ (1999) The analysis of soil erosion analysis in the watershed using GIS. Doctoral Thesis, Gang-won National University

Shrestha DP (1997) Assessment of soil erosion in the Nepalese Himalaya—a case study in Likhu Khola valley, middle mountain Region. Land Husbandry 2(1):59–80

Singh G, Babu R, Narain P, Bhusan LS, Abrol IP (1992) Soil erosion rates in India. J Soil Water Conserv 47(1):97–99

Singh PK (2000) Watershed management (design and practices). e-media publications, Udaipur

Singh AK, Kothari M, Mishra PK, Kumar V, Srinivasa Reddy GV, Purohit RC (2007) Soil erosion research under simulated conditions for black soils of Southern India—a review. Agric Rev 28(l):63–68

Staples C, Ahmed S, Ewers RM (2012) Sensitivity of GIS patterns to data resolution: a case study of forest fragmentation in New Zealand. N Z J Ecol 36(2):203–209

Subramanya K (2009) Engineering hydrology, 3rd edn. Tata McGraw-Hill Education Private Ltd., New Delhi

Suriyaprasit M, Shrestha DP (2008) Deriving land use and canopy cover factor from remote sensing and field data in inaccessible mountainous terrain for use in soil erosion modelling. The Internatonal Archivs Photogramm, Remote Sens Spatial Inf Sci 37(Part B7):1747–1750

Symeonakis E, Drake N (2004) Monitoring desertification and land degradation over sub-Saharan Africa. Intl J Remote Sens 25(3):573–592

Tamas J, Kardevan P, Kovacs E, Takacs P (2005) Evaluation of environmental risks of non-point source heavy metal contamination using DAIS sensor. In: Proceedings of 4th EARSeL workshop on imaging spectroscopy, Warsaw University, Warsaw

Tweddales SC, Eschlaeger CR, Seybold WF (2000) An improved method for spatial extrapolation of vegetative cover estimates (USLE/RUSLE C factor) using LCTA and remotely sensed data

Uddin K, Murthy MSR, Wahid SM, Matin MA (2016) Estimation of soil erosion dynamics in the Koshi Basin using GIS and remote sensing to assess priority areas for conservation. PLoS ONE 11(3):e0150494. https://doi.org/10.1371/journal.pone.0150494

Vaezi AR, Bahrami HA, Sadeghi SHR, Mahdian MH (2011) Developing a nomograph for estimating erodibility factor of calcareous soils in North West of Iran. Int J Geol 5(4):93–100

Van der Knijff J, Jone RJA, Montanarella L (2002) Soil erosion risk assessment in Italy. European Soil Bureau, Joint Research Center of European Commission

Van Remortel R, Maichle R, Hickey R (2004) Computing the RUSLE LS Factor based on array-based slope length processing of digital elevation data using a C++ executable programme. Comp Geosci 30, 9(10):1043–1053

Vicenta ML, Navas A, Machin J (2007) Identifying erosive periods by using RUSLE factors in mountain fields of the Central Spanish Pyrenees. Hydrol Earth Syst Sci Discus 4:2111–2142

Vrieling A (2007) Mapping erosion from space. Doctoral Thesis, Wageningen University.

Williams R, Berndt HD (1972) Sediment yield computed with universal equation. J Hydraulic Eng, ASCE 98(HY12):2087–2098

Wischmeier WH, Smith DD (1958) Rainfall energy and its relationship to soil loss. Trans American Geophysics Union, 39(2)

Wischmeier WH, Smith DD (1965) Predicting rainfall erosion losses from cropland east of Rocky Mountains: a guide for the selection of practices for soil and water conservation. U.S. Department of Agriculture, Agricultural Handbook, p 282

Wischmeier WH, Smith DD (1978) Predicting rainfall erosion losses—a guide to conservation planning. AH-537. U.S. Department of Agriculture, Washington (DC)

You SC (1999) Estimation of soil erosion supported by GIS—a case study in Guanji township, Taihe, Jiangxi. J Nat Resour 14(1):63–68

Young RA, Onstad CA, Bosch DD, Anderson WP (1987) AGNPS: an agriculture nonpoint source pollution model. Conservation Research Report 35, USDA/ARS, Washington (DC), USA

Yu B, Hashim GM, Eusof Z (2001) Estimating the R-factor with limited rainfall data: a case study from Peninsular Malaysia. J Soil Water Conserv 56(2):101–105

Yusof KW, Baban SMJ (2002) A preliminary attempt to develop an erosion risk map for Langkawi Island, Malaysia using the USLE, remote sensing, and geographical information system. In: GIS development proceedings

Zaeen AA (2012) Remote sensing technique to monitoring the risk of soil degradation using NDVI. Intl J Geogr Inf Syst Appl Remote Sens 3(1)

Zivotic L, Perovic V, Jaramaz D, Dordevic A, Petrovic R, Todorovic M (2012) Application of USLE, GIS, and remote sensing in the assessment of soil erosion rates in South-Eastern Serbia. Pol J Environ Stud 21(6):1929–1935

®
FSC
www.fsc.org
MIX
Papier aus verantwortungsvollen Quellen
Paper from responsible sources
FSC® C105338